USCGC MACKINAW WAGB 83

An Illustrated History of a Great Lakes Queen

An Illustrated History of a Great Lakes Queen

by Mike Fornes

Foreword by James E. Muschell

Cheboygan Tribune Printing Co.
208 N. Main St.
Cheboygan, Michigan 49721

1st Printing — October, 2005

Front & back cover photos by Neil Schultheiss
Cover photo-illustrations by Charles Borowicz.
Page designs by Dan Pavwoski

For additional copies of this book, send $35 (includes tax, postage & handling) to:
Mike Fornes – PO Box 305, Mackinaw City, MI 49701

TABLE OF CONTENTS

FOREWORD

The U.S. Coast Guard cutter Mackinaw, known as the "Queen of the Great Lakes," was first delivered to Cheboygan on a cold December day in 1944 during World War II. The ship was greeted by the citizens of Cheboygan and by the Cheboygan Daily Tribune's late News Editor Gordon Turner. Over the years Turner kept us fully informed with many human-interest stories about the Mackinaw and her crew, and kept us aware of the various reports and rumors that kept re-occurring over the years that the Mackinaw was to be decommissioned.

Every time that these reports or rumors surfaced, the residents became upset and dismayed because the presence of the Mackinaw had not only become an important economic factor for Cheboygan but had also become a family member of our community. Gordon Turner's articles in the Tribune and his public notifications of what was going on helped the residents of Cheboygan make the powers in Washington take note of the importance of maintaining the icebreaker at its Cheboygan location.

In spite of the many political interests and ramifications from other competitive communities, Cheboygan is considered to be a central key location for an icebreaker to be located. Geographically, the ship is at a perfect harbor with a natural river depth and has easy access to waters clogged by ice depths and rugged formations. The ship is close at hand to keep the Straits of Mackinac and the Great Lakes open during the winter periods important to shipping.

After 61 years with Cheboygan as its homeport, the Mackinaw has become part of our heritage and the city's residents have developed a deep appreciation for the officers and crew and their families. We consider them to be members of our community family.

Many crewmembers have married Cheboygan girls and after serving have returned and retired here. There were periods when affordable family housing and off-duty recreational facilities were difficult to find, but eventually these obstacles were overcome through the efforts of the U.S. Coast Guard. The crewmembers that did marry and settle in Cheboygan found the area perfect for raising children.

There were times when the Mackinaw crew were in port and assisted our city's fire department in fighting some very serious fires in the downtown area by providing supportive pumping facilities. The officers and crew of the Mackinaw have also participated in our Memorial Day and July 4 parades over the years, commemorating the deeds of our servicemen and women and been involved in numerous other community activities that we hold dear.

As mayor of Cheboygan, I have had a deep respect and appreciation for the Mackinaw and its crew. I am very appreciative of having had the opportunity and privilege of serving the city as mayor and as a councilman.

I am also deeply honored to have the privilege of writing the foreword for this book by Mike Fornes. It's an outstanding historical account of the Mackinaw, the officers and the crews who have served on this great ship at their homeport - here in Cheboygan.

James E. Muschell – 2005
Mayor of Cheboygan, Michigan

DEDICATION

Like countless other residents of Northern Michigan, I never met anyone who loved Cheboygan, Mich. and the icebreaker Mackinaw more than Gordon Turner.

The late news editor of the Cheboygan Daily Tribune arrived in 1927 to begin a summer job writing for the paper and wound up staying for the rest of his life. Conversations with Gordon, like his writings, were usually based on what was happening around his town.

His town.

You couldn't speak with Gordon for very long without his reporter's instinct taking over; he'd question you on whatever topic he thought you might have knowledge. His basic nature included the belief that every person he met had a story suitable for publication in his newspaper.

But he no doubt met his favorite subject for coverage when he greeted the U.S. Coast Guard cutter Mackinaw on that cold December day in 1944. In a town the size of Cheboygan, he knew the ship offered endless possibilities for newspaper stories. Human interest stories, news stories, advance stories, you name it. To Gordon Turner, the Mackinaw many times represented the most interesting things happening in Cheboygan.

It was a policy endorsed by the Tribune, whose masthead carried the mottos, "BE A BOOSTER - TALK UP YOUR OWN HOME TOWN" and "HOME PORT OF THE COAST GUARD CUTTER MACKINAW."

"He used to come over to the ship," recalled Jim Honke, commanding officer in the early 1980s, and say to me, 'Well, what are we going to do today?' and that meant he was writing a story about us whether I liked it or not. He would always convince me that things happening on board the ship were newsworthy, that whatever the Mackinaw did was important to people in Cheboygan."

His ship.

While working as a reporter for WPBN-WTOM TV 7&4 in the early 1990s, I once followed up leads that indicated the Mac was headed for decommissioning, one of several attempts to scrap the vessel. I summoned a satellite truck to the dock for a live segment on the 6 p.m. news and lined up an interview with Larry Otto, a former Mackinaw executive officer who worked in Cheboygan as a tax accountant. I tried but couldn't reach Gordon, so I left a message asking him to be at the dock at 6 p.m. for a TV interview about the ship.

When Gordon arrived, a few minutes late, we were on the makeshift set at the Millard D. Olds Memorial Moorings, minutes from going on the air. The ship was out of town.

"What's this all about?" Gordon asked while a technician clipped a microphone to his jacket. "What are we going to talk about?"

"The decommissioning of the Mackinaw," I replied. "They announced today in Washington that the ship's funding has been cut. I know you feel strongly about the Mackinaw."

Gordon Turner was clearly shocked at this news, but he was still a competitive newspaperman and didn't like being scooped.

"I can't give you an interview about that," he stammered as the crew ordered us to 'stand-by,' "You're beating me at my own story."

He was also a gentleman with class.

It *was* his story, and it was about *his* ship. But he stayed and did the interview.

It is my hope that in writing this book I can finish telling a story that Gordon Turner began writing in 1944. Many of his news accounts appear in the following pages, courtesy of the Cheboygan Daily Tribune.

Gordon would have written this book, had he lived to be nearly 100 years old.

May these recollections, photographs and stories of the Mackinaw stand as a tribute to Cheboygan – her home port – and Gordon Turner, who loved his town and his ship more than anyone I've ever known.

Mike Fornes - 2005
Cheboygan, Mich.

The Need for the Mackinaw

She's "The Queen of the Great Lakes," and has also been called "The Great White Shepherd" and even "The Great White Mother."

In military terms, she once carried the simple name of "Hull No. 188" and "WAGB-83." Later, on paper, she was briefly called the "Manitowoc."

This is the story of the U.S. Coast Guard cutter Mackinaw, the biggest, most powerful icebreaking machine the Great Lakes has ever seen. In retirement, some history is necessary to realize why the "Big Mac" was needed in the first place, and to place a perspective on the importance of this very special ship.

For many years, when winter inevitably closed in on the Great Lakes it meant the end of shipping season. Each spring in the 1920s and 1930s saw harbor ice handled by smaller icebreakers that seemed to be

A December ice blockage typical of those that stopped shipping at the Soo in the early 1900s. Nearly two dozen ships clogged the St. Mary's River in this photo, before the Mackinaw was built.

good enough to eventually open ports of commerce. However, the Straits of Mackinac and the St. Mary's River system connecting Lakes Huron and Superior were annually locked in a frozen blockade that was impassable to any vessel.

During World War II when all available escort vessels were needed in the Atlantic, the 165-foot icebreakers Escanaba and Tahoma were transferred to North Atlantic service leaving the Great Lakes without any effective icebreaking capability. This came at a time when industrialists in a war-expanded economy were demanding lengthened navigation seasons to offset the reduced building of freighters.

In addition to the war being fought in Europe and the South Pacific, factories situated along the shores of the Great Lakes were engaged in a war against time and Mother Nature. The majority of the nation's supply of iron ore is located at the upper end of Lake Superior while the majority of factories and steel mills are at the lower end of Lake Michigan and Lake Erie.

The passageways of the Great Lakes normally freeze in late December and stay that way until late March or early April, closing shipping lanes for nearly four months.

American factories that in 1941 began concentrating on building warships, tanks and planes needed every ounce of steel they could get delivered to their assembly lines. To help fuel the war effort, shipping on the Great Lakes needed to be open year-round.

The need for a longer navigation season was so great that Admiral C.A. Park, then commandant of the Coast Guard, estimated that if the new cutter could succeed in prolonging the full movement of freight vessels by even ten more days each winter it would mean three and one-half million tons of iron ore, limestone and coal moved to the blast furnaces of Great Lakes steel mills or 120 million bushels of grain moved to the lower lakes to feed American soldiers.

Many Michigan legislators stepped up over the years to help the Coast Guard protect the Great Lakes. On Dec. 17, 1941 U.S. Rep. Fred Bradley, R-Michigan, sponsored a bill to construct an icebreaker specifically designed for duty in these waters. However, the bill was opposed on the contention that ocean icebreakers could come to the Straits if necessary in an emergency.

Bradley was not satisfied.

"The ocean crushers," he related, "draw 27 feet and we have no channels that are 27 feet."

In addition, the idea of crushers breaking ice for many miles just to reach the Straits did not seem a practical solution to Bradley.

As a result, he continually fought for the bill until it was passed by the House of Representatives. Senator Prentiss M. Brown, D-Michigan, of St. Ignace sponsored a companion bill that was passed by the Senate.

Fred Bradley, the Rogers City congressman whose legislation funded a giant icebreaker.

Prentiss M. Brown was a Democratic senator from St. Ignace who backed the efforts of Bradley, a Republican, to get an appropriations bill passed in the Senate to build the Mackinaw.

The Rogers City congressman, who was on a committee overseeing the waterways, even included a restriction in the design to keep it from leaving the Great Lakes, but that detail was left to the shipyard.

Bradley secured passage of the bill and obtained funding from a special presidential fund used to expedite desirable projects.

An unusual twist of fate then led to a change of the new ship's homeport.

Each night Myrt Riggs, publisher of the Cheboygan Daily Tribune, would lord over his employees and see to it that the paper was composed, printed and readied for delivery. To say that he was a "hands-on" publisher would be putting it mildly.

Myrt Riggs, the longtime publisher of the Cheboygan Daily Tribune, wanted the Coast Guard in his town.

One night, with his work completed, Riggs happened to check the newswire one last time before departing for home. He noticed a story about the passage of the bill for a new giant icebreaker to be built for the Great Lakes, to be one day based in Milwaukee, Wis.

Riggs couldn't believe Milwaukee's good fortune in getting the ship. He thought a Coast Guard presence was exactly what was needed in Cheboygan, Mich. A legendary phone call from Riggs to Bradley's home after 2 a.m. got the ball rolling.

"We've got to have that ship here in Cheboygan," Riggs is said to have told the congressman, who sleepily agreed with the publisher's logic. "Cheboygan is a much better geographic center for an icebreaker to be based. The Straits of Mackinac is the main key to keeping the lakes open, and the St. Mary's River system is the other. That ship absolutely belongs in Cheboygan."

It is unknown what promises Riggs got from Bradley but the newspaperman kept up the pressure beginning the next day and continued with a deluge of letters, telephone calls and influential contacts.

The ship, known at this point as the Manitowoc, was soon to be destined for a port far across Lake Michigan from Milwaukee and through the Straits of Mackinac, thanks to Bradley's influence from Riggs' persuasion. A short time later, a document of registry in the University of Detroit's Marine Historical Collection states that Hull No. 188 was begun with a keel-laying March 20, 1943 at the Toledo Shipbuilding Company. But the Manitowoc's name, as well as its destination, had changed.

The ship was now to be the Mackinaw, named for Mackinaw City – her main area of operations – and would be based in Cheboygan.

This bronze dedication plaque hangs in the Mackinaw's crew's mess, listing important dates in the ship's construction and an explanation of the icebreaker's name.

Building A Great Lakes Queen

In the spring of 1943 the Toledo Shipbuilding Company began work on the new vessel, under the direction of J.W. Massenburg, the yard's superintendent who had charge of the construction.

The design of this ship was undertaken by Gibbs and Cox, naval architects, in a special icebreaker design section set up specifically for the new Great Lakes crusher and her semi-sister ships to be constructed in a classification known as the Wind Class.

Under the guidance of Cmdr. E.H. Thiele, who later retired as a Rear Admiral, all known worthwhile icebreaker design features were examined. All the best components possible were incorporated into the new ship's design. In addition, other modern shipbuilding techniques and designs were utilized as well as other equipment developed specifically for the new vessel, which would become the world's most powerful icebreaker and eventually hold that title for almost 30 years.

Completion bonds and penalty clauses are routinely written into shipbuilding contracts of this sort, and the Mackinaw was no exception. Hull No. 188 underwent some major delays and soon construction began to fall seriously behind schedule. Despite the introduction of women into the workforce, wartime personnel numbers were often low at factory or construction jobs requiring highly skilled, heavy-duty equipment operators and plain old-fashioned grunt-work laborers. Both are required when moving around tons of steel plating and machinery to build large ships. However, many women worked on various phases of the ship's construction.

In addition, America's scrap iron drives and emphasis on moving iron ore to port had become a

The Mackinaw's keel was laid on March 20, 1943 at Toledo. The ship was known as "Hull No. 188" and later became the Manitowoc, but only on paper. She became the Mackinaw while under construction.

standard way of life for industrial concerns by 1943, but many of the Mackinaw's systems and equipment were so revolutionary and unique that the vessel's ever-changing specifications slowed the building process even more. This was going to be a bigger icebreaker than had ever been built by anyone in the Toledo yard or anywhere else, incorporating many new and untried designs. The frequent delays mounted up massive costs when deadline after deadline was not met.

Soon, the job of building the Mackinaw proved to be too much for the Toledo Shipbuilding Company, and the firm declared bankruptcy. Now the unfinished hull sat in the yard, waiting for an answer as to

The 1943 construction of the Mackinaw featured closely-spaced ribs to bear its extra thick hull plates for icebreaking – 1 5/8 inches thick from the keel to just above the waterline.

whether it would ever be completed to sail one day on the Great Lakes.

To the rescue came the American Shipbuilding and Drydock Company, which agreed to complete the mammoth project. An understanding was reached with the government that the work would be performed on a no-penalty, cost-plus-zero basis. The ship was badly needed and neither the Coast Guard, the U.S. Government nor the factories along the shores of the Great Lakes were in a position to haggle over completion date details. The job would be done right, as soon as possible and the new company could start

This photo, taken in 1944, shows the Mackinaw with its completed hull. The superstructure and mast assembly were added later. The ship was built with a large fuel capacity – enough to keep going for six months.

right away.

When stepping the mast, to a height that would bc 105 feet above the waterline, workmen in the yard followed a shipbuilder's custom of placing coins under the base of the mast. Legend has it that contributions that day amounted to more than $70 in silver dollars, pre-1944 currency, that is to this day still under the mast of the ship.

The final cost of the Mackinaw was $10 million, and as launch day approached after a couple of surprisingly mild winters, skeptics and naysayers began touting the new crusher as a "white elephant." Some even referred to the Mackinaw as the "Coast Guard's Folly."

The day would soon come when the critics would eat their words.

More than 70 silver dollars were said to have been placed under the Mackinaw's mast step by yardworkers, a good luck tradition practiced by shipbuilders. The Mac's giant bow propeller is visible in the shadows under the forward portion of the bow.

Destination Cheboygan

The last place most people in Toledo, Ohio would have wanted to be on March 4, 1944 would have been on the open dock of the Toledo Shipyard, where "a driving snowstorm and savage winds" presided at the launch of the new icebeaker, the Toledo Blade reported.

The newspaper account of the event said that "During the pre-launching ceremonies, Vice-Admiral Russell R. Waesche, his wife, city officials and officials of the shipbuilding company as well as the many spectators shivered alike in the snow which struck faces and hands like buckshot."

In his speech, Waesche praised the company's workers who were responsible for construction of the ship and also cited the numerous women who had a part in building it.

The launching, originally scheduled for 12 noon, was delayed slightly by the disagreeable weather but at last Mrs. Waesche shattered a bottle of champagne on the ship's bow at 12:25 p.m. Her aim was true and the champagne provided a fitting baptism, but "the ship failed to move for what seemed like minutes to the spectators, but in reality it was seconds."

The Waesche's had a son who would one day serve aboard the ship.

As a crowd of 3,000 cheered, the 290-foot Mackinaw then slid off the launch platform and hit the water sideways, pushing a tidal wave of Maumee River water up over the pier and several small buildings at the dockside. The Mackinaw was at last in cold fresh water, much to her liking, and Toledo's pride of the Great Lakes was now afloat and soon secured to the wall.

In the following months the ship eventually underwent sea trials, and the builders began to see exactly what they had created. The Mackinaw was a magnificent, powerful vessel unlike anything the shipyard had ever built before. Three other similar, smaller designs had been constructed on the same general specifications and sent to duty on the Sea of Okhotsk, given to Russia under a lend-lease agreement. A fourth, retained by the United States, was sent for duty in Greenland.

However, the Mackinaw's particulars were designed for Great Lakes duty only.

The Coast Guard claimed the icebreaker could outlast a battleship in heavy weather. The fat, football-like shape was not obvious unless viewed from above, and a Detroit News reporter, Kendrick Kimball, referred to the Mac in a 1945 article as the "answer to a seaman's prayer," and "the most modern ship on the Great Lakes."

The final five-day trip provided a grueling test on storm-swept Lake Huron in the fall of 1944 under the watchful eyes of three Coast Guard captains who served as a trial board. District Coast Guard Engineer G.W. Cairns from Cleveland, Capt. Beckwith Jordan, district officer from St. Louis, Mo., and Capt. L.B. Olsen of the New London, Conn., Coast Guard Academy put the ship through many conditions that the Mac would routinely handle in the more than 60 years to follow.

March 4, 1944 – The Mackinaw was launched sideways into the Maumee River. Unknown to many of the shivering spectators who hurriedly filed out after the launch was the fact that three people were injured at the conclusion of the ceremonies when they were struck by a large piece of structural steel that was blown over by the wind.

The Blade listed the injured as Rodney Anderson of Milwaukee, Wis., who suffered a fractured leg; Ruth Anderson of Toledo, his sister-in-law, who sustained back injuries, and Paul Bodi of Toledo, injuries unknown. All were transported by ambulance to Toledo St. Vincent's Hospital.

Lt. Cmdr. Charles Gonyaw of Buffalo, N.Y., captain for the Great Lakes Transportation Company directed the maneuvers. He was aided by J.W. Massenburg of the Toledo Shipbuilding Company that built the Mackinaw.

The Blade reported that the vessel's tests included towing the Coast Guard cutters Acacia and Arrowhead against their full throttle pull; reduced and sustained power runs; 15 speed runs over a measured nautical mile off Sturgeon Point, Mich.; forward propeller action and immediate reversal from top speed forward to full power astern.

The trial board and Adm. Harvey Johnson, chief Coast Guard engineer, agreed after the intense trials that the Mackinaw was equal to her icebreaking assignment.

However, the vessel did not encounter ice conditions!

As a result, the crew was unable to test the ship's novel bow propeller and the combined might of its two 14-foot stern propellers, pitting 5,090 tons of metal and equipment against lake ice. But the captains who judged her agreed that she was equal to her primary task – keeping shipping lanes clear of ice in late fall and early winter, and opening them in early spring.

"She behaved well," Johnson told the Blade. "She seems fundamentally sound and should do her job effectively, a job doubly important during the war."

To the landlubber, the Mackinaw was literally a floating city. To the lake sailors who tested her, she appeared as a dream come true – a ship that would be able to win out in a slugging match with the heaviest ice the Great Lakes can offer.

Those who prowled through the vessel were impressed by its quietness due to cork insulation, its many conveniences for the crew, its huge engines and sturdy construction. The Blade noted that the Coast Guard did not even forget the comfort of the man in the crow's nest – "at all times he will be snug and warm because over him will be a covering of shatter-proof glass, while inside is an electric heater to warm his feet."

Cmdr. Edwin J. Roland

On December 20, 1944, the Mackinaw was commissioned as a Coast Guard vessel and turned over by J. Burton Ayres of the Toledo Shipbuilding Company to Cmdr. Edwin J. Roland, a veteran sailor from Buffalo, N.Y. who was named as the ship's first captain.

The Toledo Times reported that all officers and men aboard the icebreaker marched to the stern and heard a brief talk from Ayers in which he described the Mackinaw as "the finest vessel of her type." He lauded the efforts of 900 Toledo workers who participated in various phases of the production of the ship, destined to play an important role in Great Lakes commerce.

Roland accepted the ship on behalf of the Coast Guard, and the American flag was hoisted above the Mackinaw. Chaplain J.W. Quinton gave the benediction, and the Mackinaw passed over to become property of the U.S. Coast Guard.

She was headed for Cheboygan.

Opening Day

A few days after Christmas in 1944 a proud, white vessel steamed up Lake Huron, bound for her new home in Cheboygan.

Cmdr. Roland had prepared his crew of 130 officers and Coast Guardsmen for handling the cutter in the enormous task that awaited them – the breaking of ice in quantities and at locations never before accomplished. Roland knew he had a very special ship to present to the people in Cheboygan.

At the Cheboygan Coal Dock, also known then as the Olds Dock, a welcoming reception was being readied for the ship and crew, "a winter celebration that stands as one of the best one-day celebrations ever held in Cheboygan," according to a Cheboygan Daily Tribune account of the event.

The date was Saturday, Dec. 30, 1944.

At precisely 3 p.m. the Mackinaw was sighted just to the west of Fourteen-Foot Shoal Light. The weather that afternoon was "a little overcast with a low ceiling" but still ideal for the event. The ship encountered no ice in the Straits, but "as she majestically slid up the river to her berth the eight inches of ice in the Cheboygan River slid under her powerful bow as thin glass."

The icebreaker tied up just south of the City of Munising, a ferry wintering in the river. The Cheboygan Victory Band then played "Semper Paratus," with a backdrop of a huge welcoming banner displayed across the dock building. Cheboygan Mayor William Ripley was the first civilian to step aboard and presented Roland with a key to the city.

Present for the occasion, organized by local milling company owner Floyd Daugherty, were the color guards of the American Legion and Veterans of Foreign Wars, the Boys Club Fife and Drum Corps, and approximately 3,000 spectators who cheered despite the snow and cold. Dignitaries on hand included Admiral C.A. Park, Congressman Fred Bradley and State Representative Hugo A. Nelson of Indian River.

Also attending, the Tribune reported, were "Mr. I.L. Clymer, head man of the Calcite Corporation over at Rogers City and president of the Bradley Transportation Company; J.W. Massenburg of the Toledo Shipbuilding Company who was accompanying the boat with his son to this city and was met here by Mrs. Massenburg who made the trip from Toledo by train; Circuit Judge Ward I. Waller and Probate Judge O.T. McGinn; Captain Hilliard Bentgen of the Michigan State Highway Ferry Fleet and Captain F. F. Pearse of the steamer Carl D. Bradley who joined the Mackinaw at Rogers City and came to Cheboygan."

The master of ceremonies was City Alderman Steve Majestic, industrialist and war production manufacturer. Radio station WSOO of Sault Ste. Marie strung lines to the dock to broadcast the proceedings and many newsmen and photographers were at the scene. Cheboygan had no radio station then.

Following the welcoming reception, Roland invited the public to board the vessel for a tour and many citizens accepted for two hours. Then the entire throng gathered to begin a parade down Main Street and doubled back, passing the approving crowds lining the street and ending at the Gold Front banquet room. The guards were boys who shouldered wooden rifles.

Saturday, December 30, 1944 – The day the Mackinaw arrived in Cheboygan, Mich. The ship tied up at the Coal Dock, its temporary home on the west side of the Cheboygan River until moving to the turning basin some five years later. A crowd of 3,000 people greeted the new icebreaker.

A dinner was set up for 500 people, with the Mackinaw officers and crew as guests along with any servicemen who happened to be in the area. The meal began promptly at 6:30 p.m., with fifty young women serving at once.

For the general public, Marvin McLelland and his Gold Front staff had arranged a $2-a-plate affair, featuring tomato juice cocktail, celery hearts, olives and relishes, roast sirloin of beef, garden fresh wax beans, creamed whipped potatoes, chef bowl salad, Parker House rolls, orange pineapple ice cream and coffee.

"An absence of cigarette smoke was noted," stated Gordon Turner in his newspaper coverage.

The music for the dinner hour and dance following was played by Earl Thomas and his 15-piece orchestra from Sault Ste. Marie. Master of ceremonies at the dinner was Cy O'Toole and the invocation was offered by the Rev. Edmund Mantei. "America" was sung by the entire assembly, led by Spencer Jones, then the St. Charles Glee Club sang "Semper Paratus" and an originally composed song called "Salute to the Coast Guard" welcoming the officers and men of the Mackinaw to Cheboygan, under the direction of the Rev. George A. Gougeon.

The Tribune's coverage of the event states that communities represented at the dinner included Sault Ste. Marie, Harbor Springs, Onaway, Petoskey, St. Ignace, Rogers City, Alpena and Mackinaw City. Several of the cities sent their mayor, city manager, alderman or other high officials. The Hammond Bay Coast Guard Station sent a delegate as did the Coast Guard station from Bois Blanc Island.

The Finance Committee of the group that welcomed the Mackinaw presented Roland

Cheboygan, Michigan, December 30, 1944

3 p. m. Arrival of MACKINAW. Reception Celebration at Olds' Dock
Celebration General Chairman, Floyd Daugherty

BANQUET PROGRAM

7:00 P. M.

Gold Front Ball Room
Master of ceremonies, C. J. O'Toole
Singing of "America" by entire assembly, led by Spencer Jones
Invocation, Rev. Ed. Mantei, Pastor St. Thomas Lutheran Church
Introduction, Judge Ward I Waller
"Semper Paratus" by the St. Charles Girls' Glee Club, the Very Rev. Fr. George A. Gougeon, accompanist
Talk by Admiral C. A. Park, Chief of Operations, USCG
Introduction of Officers and Guests by Judge O. T. McGinn
Talk by Capt. Hillard Bentgen, Commodore State Ferry Fleet
Talk by Congressman Fred Bradley, 11th District of Michigan
Presentation of Welfare and Recreation Check by Steve Majestic
Response by Commander E. J. Roland, commander of the Mackinaw
Close, singing by all of "Star Spangled Banner."
Music for dinner hour and dance
Earl Thomas Orchestra of Sault Ste. Marie

MENU

Tomato Juice Cocktail

Celery Hearts Olives and Relishes

Roast Sirloin of Beef

Garden Fresh Wax Beans Creamed Whipped Potatoes

Chef Bowl Salad

Parker House Rolls

Orange Pineapple Ice Cream

Coffee

Cigarettes ? ? ?

with a generous contribution "of well over $1,000" for the ship's Recreation Fund.

Roland stepped to the microphone and addressed the audience, saying, "We of the ice crusher Mackinaw count ourselves particularly fortunate that the ship on which we are to serve will have for its home port the city of Cheboygan. You people of Cheboygan have been generous beyond all just expectations in your reception of my officers and men and have established a bond between this ship and the city that will exist no matter how far distant from our homeport our operations may take us."

He continued, "I speak for all the officers and men aboard when I say we have a very deep, a sincere and personal gratitude to the people of Cheboygan for their gift which will make possible all those facilities so vital to fighting men at sea. Each officer and man, many in foreign waters, has received intensive training in every phase of their respective duties. Each can be counted upon to perform his duties fully, efficiently and dependably.

"We are ready for our new work," Roland concluded. "We will accomplish what we are dedicated to do. We will, just as you of Cheboygan are doing, contribute to victory over the enemies of our country and over the hazards to our Great Lakes shipping — not in gestures, but in our deeds."

The Mackinaw truly belonged to Cheboygan after that day, but the celebration was brief. Three days later the ship was headed out of the river for its first official duties.

This 1958 photo shows the Mackinaw with a coating of ice on her bow – a common sight after a frozen Great Lakes gale.

Establishing the Icebreaking Business

Just three days after the massive celebration to welcome the Mackinaw to Cheboygan, the ship departed on its first mission. A ten-day trip was ordered, summoning the Mac to Sault Ste. Marie, Bay City and Chicago to perform icebreaking and escort duties. Everybody wanted the ship to help them out of trouble.

At the Soo it took over for the icebreaker Chapparal, stuck in heavy ice in a narrow channel of the St. Mary's River. After freeing the Chapparal, the Mackinaw headed for Bay City to open a shipping channel there. The next stop was Chicago, where the giant icebreaker gave escort service to several newly-constructed naval craft en-route to ocean duty.

The Mackinaw was not designed to break ice by ramming or plowing into ice fields. Instead, she rides up on top of the ice and the sheer weight of the ship crushes through the ice. The vessel also has a heeling and trimming system that allows 160 tons of ballast water to move from tanks on one side of the ship to another in 90 seconds, or a flow rate of 54,400 gallons per minute. This action produces an arc of 24 degrees, allowing the ship to rock back and forth and forward and aft. Rocking the ship pushes the ice away from the hull so the vessel is able to break out of the ice jams that stop the forward motion of the ship.

An early ice breaking photo, probably around 1950. Already crewmembers were enjoying walking around the ship on the ice.

USCGC Mackinaw breaking ice in the Straits of Mackinac during the 1950 spring breakout.

"One problem that developed with that system," recalled Charles Dorian, a Mac crewman from the 1960-era, "was that excess water could sometimes come out the vents onto the deck during the shifting of the water. Naturally, it froze right away, and walking the decks could be a bit hazardous."

Navy engineers designed a bow-propeller system as an improvement upon the battering-ram technique used by more conventional icebreakers. The front propeller's rotation draws water from beneath the ice from 150 feet in front of the ship, causing the ice to weaken and the ship to crush it like a hollow shell as the two rear propellers drive the reinforced bow over the top of the floe. The "wash" or bubbling action created by the bow propeller also acts as a lubricant to reduce friction on the hull.

The "Mighty Mac's" diesel-electric power plant is powered by six Fairbanks Morse ten-cylinder engines that can produce 2,000 horsepower, allowing the ship to travel at a speed of 18.7 knots or 21.6 miles-per-hour. The engines are directly connected to six electric generators. The power produced by these generators can be used in various combinations to drive one, two or three main propulsion motors of 5,000 horsepower each, which turn the two 14-foot stern propellers and the 12-foot bow propeller. The stern propellers, made of cast steel, weigh 10.7 tons each and the bow propeller alone weighs 7.2 tons!

The Mackinaw is capable of breaking 38-40 feet of "windrow" ice and 42 inches of "blue" ice. Great Lakes ice can take a variety of forms and consistencies:

• Pack ice is solid, unbroken ice that can stretch for miles and is often blue in color, typically found in

"Pancake" ice in the Straits of Mackinac near Mackinac Island.

the Straits of Mackinac. It is the toughest to get through and the result of the coldest weather the area has to offer.

• Basic or "new" ice takes the form of sheets, but by the action of wind and weather is transformed into different obstacles of all types, often piling up into massive windrows.

• Brash ice is sheet ice broken up or crushed like a giant Sno-Cone, with mushy, granular consistency.

• Pancake ice consists of round cakes one or two feet thick and varying in size, usually all compacted together in a large floating mass.

• Floe ice is made up of large pieces of sheet-ice floating freely. Unbroken floes are usually all blue ice, but as floes break away from shore they grind together and compact themselves, causing pressure. The upward pressure causes ridges, but the downward pressure builds what are, in effect, small icebergs just beneath the ridges. About one-eighth of the berg is represented in the ridge visible on top of the floe, while seven-eighths is usually below. It is here, in the pressure ridges, where icebreakers encounter the greatest difficulty.

The Mackinaw attacks the ice by getting up on it and shearing or folding it downward. This means that a ship following her through pack ice has to keep close at all times, so that the ice does not close in behind the icebreaker before the other vessel passes clear. Pressure ridges can create a real hazard for a vessel attempting to follow an icebreaker, because the Mackinaw can be stopped very suddenly and be overtaken by the following ship. A collision would seem inevitable, but seasoned commanders have learned to avert that hazard.

When the Mac hits a stubborn pressure ridge, her main propellers are run at "full speed ahead" so that

A Lake Michigan gale piled four to five inches of ice on the Mackinaw's deck in these late 1960s photos.

Wind-driven spray could build up a thick coating of ice on the Mackinaw's deck.

the prop wash causes the trailing ship to either stop or shear to the right or the left of the icebreaker's stern. The technique usually works well.

Capt. Joseph Howe said the key to the job is the strength of the ship.

"Brute force – just power – that's the way to break ice," Howe told United Press International for a story. "This ship was built well for a job it does efficiently."

"It was a lot like a turtle race," said Frank Umbrino, commenting in 2004 on icebreaking operations 34 years after he left the ship. "Those are my favorite memories, though. We would get six engines on line and go maybe one knot, a knot and a half, and then we would get hit by some big storms after we broke the ice that would just knock the boat around. The thickness of the ice and progress we didn't make through the ice, that was the big thing."

Umbrino said the ship took a 48-degree roll in a storm coming out of Benton Harbor during the winter of 1968.

"I wasn't sure we were coming out of that one," he remarked.

Joe Etienne said the worst ice he ever saw was in Lake Superior, but the Straits of Mackinac would get bad too.

"Sometimes the car ferries would get stuck and we'd have to break them out too, just like the freighters," Etienne remembered.

As the executive officer on another cutter, Etienne once saw the propeller blades break when the order came to throw the engines into reverse. A prop blade broke on the Mackinaw the same way in the St. Mary's River. The ship had to go into drydock for repairs. It is quite an operation when the stern propellers are 14 feet high.

"The Mackinaw is a great ship and very unique," stated Fred Thurston, a chief boatswain's mate from 1969 through 1973. "That was during the Vietnam War and we used to lead a convoy of 15 iron ore carriers through the ice from Duluth to Buffalo and then turn around and go back again. I'm from Chattanooga and that was cold, believe me."

The Mackinaw is also fitted with a reinforced notch in her stern so that following ships can nudge right up against her and follow the Mac through light ice, but that is not practical where there are pressure ridges.

Running in ice and running in open water are two entirely different operations and the Mackinaw handles like two different ships under those conditions. It requires very specialized skills from a wheelsman to navigate the ship under either circumstance.

Another feature of this magnificent vessel is a very large, constant-tension towing winch used to tow disabled vessels through ice fields. Because of the

Cutting a circular path around the ore-carrier Mercury in the late 1940s.

danger of being rammed from behind, the bow of the vessel being towed is placed into the stern notch before the towing operation is begun.

Howe said "We can and do get within 10 feet of a stuck vessel, which is close quarters when you're dealing with two ships. You get up in front of a tanker that's stuck and then you just pour the juice to her and they're on their way to freedom."

Although the danger of a collision is removed through this procedure, the Mackinaw's steering ability is greatly compromised and this method of towing is restricted to large bodies of water and favorable weather conditions. Some crewmembers say they served their entire duty of several years aboard the Mackinaw without ever using this procedure, while others say they recall many towing episodes.

"Sometimes we would run only two engines while breaking ice to minimize the number of crew we needed working down there," Etienne said. "Then we would end up getting stuck in a windrow of ice. When you've got a ship coming up on the stern end in the channel behind you and you are stuck with only two engines on the line, you've got something to be concerned about."

Etienne said he recalls the bow propeller being seldom used, only when an especially large windrow confronted them.

The ship is 290 feet long and 74.5 feet wide. Her hull plating is one and five-eighths inches thick from the keel all the way to a few feet above the waterline. The frames are very sturdy, deep and closely spaced. The Mackinaw is built to handle anything the Great Lakes can throw at her!

Notching up the Mercury for a tow.

"This boat is the most remarkable ship I've ever been on," said Byron Trerice, crewman from the early 1950s. "Blue ice was the toughest thing you'd see, and we never found blue ice that we didn't conquer. In Whitefish Bay we ran a steam hose down 19 feet to see how thick it was, and we broke that windrowed blue ice apart."

It did not take long for the ship to establish herself as the Guardian Angel of the Great Lakes. An especially long and cold winter hit the region in 1946-47, and the Mackinaw used the opportunity to really show her stuff. The war was over, but from March 25 until May 21 the icebreaker provided direct assistance to more than 1,500 vessels flying flags of the United States and Canada, and hundreds of other ships were able to use channels opened by the Mac. This tradition was to be carried on many times throughout future years of service, proving that nobody can do it better than the Mackinaw.

As an example, during a span of the years 2000 through 2003 – with fewer, but bigger ships working the lakes – the Mackinaw was deployed for more than 200 days of icebreaking and assisted more than 245 vessels throughout the Great Lakes.

Leo Cocciarelli's photo of the William P. Palmer, notched up for a tow through thick ice in 1952. Cocciarelli said the Mackinaw was stopped by windrowed ice, causing the Palmer to collide with the Mac.

The Palmer cut through the Mackinaw's stern, sending the icebreaker to dry dock for repairs. "We used to train for this sort of emergency all the time," said Cocciarelli, "and then when it happened, we ran for our cameras to get a picture of it."

The Mackinaw has crushed its way through the worst ice conditions the Great Lakes can dish out for over 60 years. Note the helicopter on the fantail deck in this 1950 photo.

The winter of 1984 created an unusual situation when tons of ice jammed up the St. Clair River near Algonac, Mich., blocking the passageway from Lake Huron through the area of Port Huron, Mich., and Sarnia, Ont. More than 50 ships waited out the conditions while the Mackinaw worked with three other Coast Guard cutters and two Canadian vessels to break the worst ice jam in decades.

The Mac left Cheboygan April 9 that year for what was expected to be a six-to-10-day stint clearing ice. It turned into a grueling marathon operation that lasted nearly a month.

"It was an outstanding example of cooperation," Capt. Paul Taylor told the Port Huron Times Herald, complimenting the crew of the Canadian icebreaker Des Groseilliers. "I never faced a situation like this in my life - even with my Arctic experience."

Shippers lost about $1.7 million each day their freighters were forced to remain at anchor, the Cleveland-based Lake Carriers Association estimated.

Tom Seigo, of Caro, Mich., told the newspaper he would gladly do it all over again.

"I feel very fortunate to be on the Mackinaw," the machinist's technician said. "Helping ships on the Great Lakes is what I wanted when I joined the Coast Guard. As long as I know it's going to be appreciated, I don't mind at all."

Also seeking the brighter side was the ship's executive officer, John Bannon, who said he'd never encountered a situation like the St. Clair River ice jam of 1984 despite serving on two polar icebreakers.

"It's not the kind of thing you'd want to do every day, but it was interesting."

Bannon praised the crew, especially the men in the engine room who kept the ship's six large engines in service around the clock.

"That's a real credit to the troops down in the hole that nobody ever sees and nobody ever thinks about," he said.

Machinist's technician Pat Milligan wasn't sorry to have the work behind him.

"The only time I ever want to see ice like this again is on a snow cone," he said.

The ship's public affairs officer, Ensign Jim Meador, said it was actually enjoyable.

"Some guys like to look at women. I like to look at ice. It's beautiful," Meador told the Detroit Free Press. "God, don't put that in the paper!"

Laying a track in the St. Mary's River channel with two lakers in pursuit during "Operation Taconite", 2003.

A helicopter was often used on the ship in the early years of service.

1956 - A Coast Guard helicopter departs the Mackinaw's fantail deck for a transport, a medical evacuation or to check ice conditions ahead of the ship.

Big Mac, upbound in the South Channel of the Straits of Mackinac in 1955. Bois Blanc Island is visible in the background. Note the enclosed watch station at the crow's nest on the ship's mast. It was later removed.

Lt. j.g. Jim Grabb gave the Mac a push from thick ice floes in 1955.

A 1956 aerial view of the Mackinaw clearing ice floes in the Straits of Mackinac during spring breakout.

Breaking ice in Munuscong Lake, 1955.

Approaching Mackinac Island Light.

A mid-1940s towing job for the Huron Cement carrier S.T. Crapo through pack ice.

Of Captains Courageous

Captains of the USCG Cutter Mackinaw

Cmdr. Edwin J. Roland ....(1944-46)

Cmdr. Carl H. Stober ....(1946-47)

Capt. Harold J. Doebler ....(1947-49)

Capt. Carl G. Bowman ....(1949-50)

Capt. Dwight H. Dexter ....(1950-52)

Cmdr. Willard J. Smith ....(1952-54)

Capt. Clifford R. MacLean ....(1954-56)

Capt. Evor S. Kerr ....(1956-58)

Capt. John P. German ....(1958-60)

Capt. Joseph Howe ....(1960-62)

Capt. Benjamin Chiswell, Jr. ....(1962-64)

Capt. George H. Lawrence ....(1964-66)

Capt. George D. Winstein ....(1966-68)

Capt. Otto F. Unsinn ....(1968-70)

Capt. Lilbourn A. Pharris, Jr. ....(1970-72)

Capt. John H. Bruce ....(1972-74)

Capt. Lawrence A. White ....(1974-76)

Capt. Donald D. Garnett ....(1976-78)

Capt. Gordon D. Hall ....(1978-80)

Capt. Francis J. Honke ....(1980-83)

Capt. P.R. Taylor ....(1983-85)

Capt. A. H. Litteken, Jr. ....(1985-88)

Capt. J. J. McQueeney ....(1988-89)

Capt. A. H. Litteken, Jr. ....(1989-89)

Capt. R. J. Parsons ....(1989-92)

Capt. C. A. Swedberg ....(1992-95)

Cmdr. K. R. Colwell ....(1995-98)

Cmdr. E. Sinclair ....(1998-2000)

Cmdr. J. H. Nickerson ....(2000-2003)

Cmdr. Joseph C. McGuiness ....(2003-2006)

The list of commanding officers who skippered the U.S. Coast Guard cutter Mackinaw comprises a roster of truly remarkable men. Several went on to become admirals in the Coast Guard. Many were successful businessmen in civilian life and great family men. Others retired from their service and enjoyed restful days that were well deserved. All share a distinguished place in a chapter of Coast Guard history on the Great Lakes.

Cmdr. Edwin Roland, the Mackinaw's original skipper.

Cmdr. Edwin Roland set a fabulous example for captains to follow with his immediate acceptance of the Cheboygan community, his contagious enthusiasm for building crew morale, and his ascent in a Coast Guard career that saw him reach the rank of Admiral.

He no doubt inspired his own son, Lt. Edwin J. Roland, who served as a student engineer on the ship and quipped, "It's a fine ship to learn on. You come up against things you wouldn't find elsewhere."

Others who followed added their own distinctive marks to the Mackinaw's command.

Capt. Dwight Dexter was in command from 1950 to 1952, and also served as the secretary/manager of the Cheboygan Chamber of Commerce for a time. While he assisted the Chamber, Dexter developed the Mackinaw Loan Fund for Coast Guardsmen or their families needing financial help. During the winter of 1959-1960, a drive under the general chairmanship of Bill Ripley, Richard Hess, Harold Lorton, Art Baker and Mrs. Clarence Land raised a total of $1,347.44 as a "friendly helping hand" to ease the temporary emergencies that can arise for a serviceman. Ripley

Capt. Dwight Dexter and Vice President Alben Barkley, who served with President Harry Truman.

was the mayor of Cheboygan when the Mackinaw came to town in 1944.

Many Cheboygan businesses, civic groups and private citizens contributed to the fund, available to the men of the Mackinaw.

By 1962, Dexter was an admiral, and reported that 111 loans had been made in the aggregate amount of $2,032. The fund continued to grow, and at the time $1,263 remained for loans. Dexter's foresight saw to a need that helped boost Mackinaw crewmen in the community.

It is very clear when talking to former crewmembers that opinions differ as to what makes a particular captain liked or unliked by the crew. One man's favorite captain was the next man's nemesis on board the ship.

"There are two types of captains," recalled Frank Umbrino, who sailed on the Mackinaw in the late 1960s. "There's an officers' captain and there's a crew's captain. The first type stayed in his office or his wardroom all day and didn't mingle with the crew or ever visit the guys in the engine room. The second type did the opposite and didn't require all the formalities. Just call him 'captain' and you didn't need to worry about the salute, that type of thing."

"I can remember some things being a little too loose for my liking, but I was an old school type of guy," said John Sawicki, who served as a go-between as a master chief of the boat. Chiefs were bound to think that way, he added, as they found a disciplined crew much easier to deal with.

Some captains were known for their ability to get the crew out in the community and pitch in on fundraisers, projects and worthy causes. Others helped to organize morale-building events and enlisted the crew in local competitions like bowling leagues, softball or basketball leagues. For a time the Mackinaw had its own bowling league, formed entirely of crewmembers from the ship.

Still other captains worked hard for the families of crewmembers, often through the efforts of their own wives, to organize the wives into a group that would welcome newcomers, take on service projects and see new families through the crew's long absences from home.

Each captain had his own distinctive style in leading the ship's crew through icebreaking season, cruising season, maintenance work or even shipyard refits. The topic of how the captains handled their ships and their crews is one that will surface without fail during any reunion of Mackinaw alumni.

Capt. Robert Parsons took command in 1992 and achieved a distinction for the Mackinaw that makes it even more unique. He learned that a Coast Guard ship that is home-ported at the same place for the longest time has the privilege of having the name of the port painted on the vessel. He invoked this tradition, and naturally Gordon Turner of the Cheboygan Daily Tribune got behind the cause. U.S. Rep. Bob Davis, D-Mich., also pushed for the addition and as a result the name, "Cheboygan, Mi." was painted on the stern of the Mackinaw. It's an honor seen nowhere else in the Coast Guard fleet.

Capt. Carl G. Bowman

Capt. Clifford R. MacLean

Capt. Evor S. Kerr

Capt. Joseph Howe

Capt. Benjamin Chiswell

Capt. Donald Garnett

Capt. George Winstein with Ed Pyrzynski

Capt. John H. Bruce

Capt. Otto Unsinn

Cmdr. Jonathan Nickerson

Cmdr. Joe McGuiness

Admirals W.J. Smith and E.J. Roland on board in 1966. Both served as Mac captains earlier in their carrers.

Cmdr. Harold Doebler (right) with Swedish naval officers who came to the Great Lakes to survey icebreaking techniques.

Seven former commanding officers of the Mackinaw were together at the 1974 Alumni Reunion. From left – Lilbourne A Pharris, Jr., Clifford R. MacLean, Willard R. Smith, Lawrence A. White, Edwin J. Roland, Jr., Benjamin M. Chiswell, Jr., and George D. Winstein.

Adm. Arnold Danielson salutes the 1976 Change of Command to Capt. Donald Garnett.

Capt. C.R. MacLean on the bridge in 1955.

Admiral Henry H. Bell cuts the ribbon with Capt. Jim Honke (right) at the 1982 Coast Guard Housing dedication.

Capt. Robert Parsons holds a paint can for U.S. Rep. Bob Davis, painting the homeport of Cheboygan on the Mackinaw's stern in 1991.

Above - Cmdr. Joseph C. McGuiness, left, at his change of command ceremony succeeding Cmdr. Jonathan Nickerson, at right. Rear Adm. Ronald Silva, center, presided at the affair

Right - An officer's tour of duty, like a crewman's, usually changed every two years. There was always someone new arriving or departing the ship.

Ever the gracious hosts, the Mackinaw's crew for years sponsored an annual luncheon in Cleveland on the deck for the Lake Carriers Association, a powerful ally in lobbying to keep the ship alive in its icebreaking duties. Note the Filipino waitstaff standing at left. Filipino men were employed on the ship from for many years as stewards and food servers. The practice was discontinued in later years.

The Early Crews

It would be safe to say that most Coast Guardsmen assigned to the Mackinaw in the early days of the ship's presence in Cheboygan had never heard of the city, let alone been able to specify where it was in Michigan. Some undoubtedly assumed the town was Sheboygan, Wis., but recruits today have information readily available about the location and other pertinent facts at hand for their transfer.

An average tour of duty aboard the Mackinaw has always been around two years in duration. For some, especially those unaccustomed to cold weather, a hitch on the Mackinaw may well represent the longest two years of a person's life. For others, those two years represent a time period that changed their lives forever and determined where they would meet and marry a spouse, have children and settle into civilian life.

When Gene Snider reported for duty aboard the Mackinaw in 1969, he arrived in Cheboygan on a dismal, rainy May night.

"I came here from Akron, Ohio and all I could see were the coal piles along the river and the lights on the sign of the Gold Front Bar on Main Street," Snider recalled. "I called home and told my folks I had no idea where I was."

Ed Pyrzynski arrived in 1946 from Chicago, and

1947 - Ed Pyrzynski served three tours of duty aboard the Mackinaw, married a Cheboygan girl and retired in Cheboygan.

1947 crew, with mascot Jack LaLonde (at right).

came to Cheboygan by train.

"The Chief picked me up about 7:45 in the morning and drove me through town to the ship, which was tied up to the Coal Dock at the time," Pyrzynski reminisced. "I said to my driver, 'where is the city?' and he told me 'you just went through it.' I remember asking myself what I was in for."

Both men married in Cheboygan and made homes in the area.

"The town grew on me," Snider explained. "I'd always lived around big cities. I came up here and liked it. Cheboygan is quiet and peaceful and I've been here ever since."

Things were very different during that era, with

Chiefs of the Mackinaw — 1956

the ship moored on the west bank of the Cheboygan River at the Coal Dock, whereas in later years the Millard D. Olds Memorial Moorings were developed on the river's east side, next to the turning basin where the ship also docked for many years. In 2005, the site was again upgraded for use by two vessels and a station was added for buoy storage and repair.

The original location was railroad accessible, there was plenty of room for parking, the crew could walk to town in five minutes and the first business they came to was the Gaiety Bar, a favorite watering hole with a bowling alley. The Club 27, known as the "Bucket of Blood" was also close by. There was LaLonde's Inn, Snoopy's, the Gold Front, and the Charley Bar. There was the Silver Bar, Johnnie's, the Beaver, Guyette's and the Fiesta. Some of these names were interchangeable at the same locations and would depend upon the era to identify what a particular bar was called at a particular time. Sailors with transportation sometimes went farther away to other taverns like the Evergreen, the Two-Mile Inn, or even Buck's Tavern. Most stayed downtown.

The Coal Dock was also a dirty place. Coal dust got on the ship, in the ship and everywhere else when

The Mackinaw moored at the Coal Dock on the west side of the Cheboygan River in the early years. Note the coal piles at right. The ship's Jeep, parked at left near the bow, was routinely lifted aboard by a deck crane for use in other ports. The Jeep was once driven off the edge of the pier into the river by an unwary crewman. The crane lifted it out.

Jim Kingeter enjoys a pastie on board in 1957. Bill Lorenz's mother used to make them for the crew when the ship would pass through the Houghton - Hancock area.

1947 Mac radioman at work in the radio shack.

Capt. MacLean presents the 1955 bowling league trophy to Joseph Kolanko.

1950 aerial view of the Cheboygan River mouth. The Mackinaw had moved across the river to the turning basin from the Coal Dock, its first home.

the wind would blow dust towards the air intake vents. Parked cars, uniforms, even food on board could have the presence of coal dust some of the time. If you set a bag down in the parking lot, it immediately had coal dust on it. There were coal piles very close to the ship.

David Kaplan came to Cheboygan in 1948 by train at the age of 17. He reported with four others from boot camp at Mayport, Fla. He married his wife, Beatrice McCoy, whom he met in Cheboygan and they had six children. He served until 1953.

"They used to pay us in cash instead of by check in those days," Kaplan said, "and they would pay us in $2 bills to show the town how much we spent here. I was paid $50 a month when I came aboard. When we got married we rented a house on Western Avenue for $25 a month — furnished — and it had an outhouse."

This 1955 deck inspection featured a number of different hat styles among the Mackinaw's crewmen.

Renee Zimmer came aboard in November of 1968 and remembered the ship being called out the very next day to aid a freighter that had run aground.

"I was an E3, a seaman, but I'd been on the beach for a year," Zimmer said at the 2004 Alumni Reunion. "Everyone else on deck were E2's, so I guess they thought I knew something. I was told to get the shot line ready and to pass a bowline knot through it. I hadn't seen a bowline since boot camp, but I did it and wouldn't you know a minute or so later they said the line had parted. I figured I was in real trouble, but fortunately it wasn't my knot. That's what I'll always remember about my first experience on the Mac."

Everyone who ever sailed on the Mackinaw remembers funny things, sad moments, excitement, boredom, friendships and squabbles here and there among crewmembers.

"Sometimes you had to let them work things out amongst themselves," reasoned Captain Jim Honke, who served as the commanding officer of the icebreaker from 1979-1982. "You want to lead them, but you don't want to be a babysitter in the day-to-day things that go on."

Honke and Pyrzynski also recalled a sailor who didn't get along with a lieutenant, junior grade, over some things.

"It was basically a power struggle," Pyrzynski said of the officer. "He used to stand guard on the

An impromtu jam session by some of the Mackinaw's musicians. The ship's band entertained in all types of weather on the fantail deck when time permitted.

quarterdeck and give guys trouble when they were coming home. The sailor poured five pounds of sugar into the gas tank of the officer's Volvo station wagon and ruined the motor."

Pyrzynski said that there were 135 enlisted men on board in those days, the early 1970s, and not one would tell who did it although most knew who it was.

"The lieutenant announced a restriction of everyone on board and wouldn't let them leave the ship," Pyrzynski said. "The culprit confessed, saying, 'I couldn't do that to you guys,' and the restriction was lifted."

The Mackinaw was a ship that sailors often requested again later in their careers if a spot was available. Many crewmen did more than one tour of duty at Cheboygan. Pyrzynski and Bill Tomac may have set the record with three. When Pyrzynski retired in 1966, the ceremony rivaled that of a change of command. Pyrzynski had been around for so long

Executive Officer Bernard R. Henry (left) conns the ship into port.

1971 crew, prior to shaving their icebreaking beards.

that he was considered a part of the ship. His family attended the retirement ceremony and will never forget that proud day in Ed's career.

"I actually cried when I walked across the Mackinaw's brow for the last time," said John Sawicki, who also served aboard several other cutters. "It was the best duty I ever had. I missed the Mackinaw tremendously as soon as I left. I should have stayed in the Coast Guard longer than I did, for sure."

PLANK OWNERS CERTIFICATE

Know Ye all men by these Presents

that William E. Henry, S.C.2c

is a CHARTER MEMBER of the

U. S. C. G. C. MACKINAW

by virtue of assignment as a member of the Commissioning Crew this 20th day of December, 1944, while serving in the United States Coast Guard in time of War.

H. A. White
H. A. WHITE, Lt. Comdr, U. S. C. G.
Executive Officer

E. J. Roland
E. J. ROLAND, Comdr, U. S. C. G.
Commanding Officer

Original crewmen received a Plank Owner's Certificate, signifying them as the first to serve on the Mackinaw. This one belonged to William Henry.

The dock at the turning basin in 1960. Quonset huts, pictured lower left, were assembled to provide onshore work spaces & storage.

A 1967 view of the same dock, with upgraded electrical service and improved brow platform. The parking area had been paved.

1956 – Bill Lorenz in the Mackinaw's radio shack.

Ed Pyrzynski sounding the Bosun's Mate whistle in the Mac's early days

Mackinaw crewmen preparing for a snowball fight while off the ship for R & R during 1947 icebreaking.

Family Life At Cheboygan

Every Coast Guardsman who ever reported for duty at Cheboygan will tell you that it is a unique town – in both negative and positive ways. With a population of around 7,500 in Cheboygan County, the city embodies small town Northern Michigan life and habits. Single crewmembers aboard the ship will tell you there's very little, if any, nightlife. Married crew with families often found the area perfect for raising their children. The lack of available, affordable housing was a problem for many years.

Like any romance, the Mackinaw's relationship with Cheboygan had its ups and downs. Different commanding officers say that the varying personalities of the crew dictated how well they got along in town. Crewmembers from different eras point to the CO's leadership as setting the tone for how the crew was perceived in the community. Most would agree there is some truth in each faction's role in getting along in town.

Jack Eckert served aboard the Mackinaw in the 1950s and contributed this story about his arrival in Cheboygan and adjustment to life as a Mackinaw crewmember just months after the birth of his first child, a son named Randy.

"Out of the clear blue in January, 1956, I received a phone call on the Milwaukee Breakwater Light Station advising of my transfer to the Mackinaw," Eckert wrote in his column, "Jack's Joint," on the www.jacksjoint.com Web site.

Jack Eckert in 1948

"We owed big time for him (Randy). Those were the days when we got free dental care from the mobile dental unit but had to pay for our babies. There were no family health benefits. With my meager EN3 pay, a small quarters allowance, plus $77.10 a month subsistence money received because I was on duty at the lighthouse, we were at the break-even point with our finances only because I was able to divert $50 of the subsistence allowance to cover our food.

"This was my first transfer with a wife and new baby and came at a difficult time because there were a lot of personal arrangements to make rapidly. We had to give away our German shepherd dog, breaking my wife's heart. We didn't have a very good car, a 1952 Ford, and were concerned

Mackinaw at the end of World War II.

about driving through the Upper Peninsula to St. Ignace and then taking the ferry to Mackinaw City, so we opted to take the car ferry from Milwaukee to Ludington. We drove aboard the City of Midland just before 11 p.m., settled down on the benches in the main cabin, and waited for time to pass as we crossed Lake Michigan. The trip was a rough ride, even for a large car ferry, but we made it without incident; we arrived tired and sleepy at six in the morning and drove off the boat, heading to Cheboygan by following the road map as best as we could.

"About 40 miles out of Ludington the radiator boiled over. I filled it and we proceeded on. About 25 miles later it happened again. Somehow we limped into Cheboygan by noon and looked around the city for a place to stay. By pure luck we found Mrs. Johnson's boarding house on Dresser Street. The rent was $5 per night. The room was rather small, but it had a big, comfortable clean bed in it. We bought groceries only for Joana and Randy. My plan was to report in and eat on the ship until we got squared away, financially and otherwise.

"Upon reporting, I found we were to sail in two days. Fortunately it was for just a short trip. Joana took it rather stoically and went about searching for a place for us to live. As luck would have it she found half of an old side-by-side duplex a few doors from the boarding house. The rent was more than we were used to paying, but we took it anyway. We figured the furniture would be in Cheboygan by the time the ship got back and we could move in. It took twelve days to arrive. Somehow the movers sent it to Texas and lost it.

"Meanwhile, we continued to live at Mrs. Johnson's. The ship was scheduled to be in port until mid-March, so there was no worry. But after the furniture finally arrived and with the apartment a mess from unpacking, the ship was ordered out to go to Midland, Ont., Canada in Georgian Bay to break out their harbor.

"When the ship got back Joana had the apartment in good shape, and we enjoyed our first home-cooked meal together in several weeks. I didn't have the heart to tell her that we had pork chops for lunch on the ship. Hers were better.

"Cheboygan in the winter is a cold place and

The famous Coast Guard "slash" was added to the ship's insigia in 1967.

does not suffer for lack of snow. A space heater on the street floor was supposed to heat the bedrooms upstairs. Randy slept in the living room with us, where we lived until the weather warmed up enough in June to move upstairs. The heating bill was more than the rent!

"The 'Great White Mother' sailed for the icebreaking season of about six weeks while Joana spent a lot of time at the public library and at church. We had just barely begun to get ourselves acquainted socially. There were a number of young couples on the ship in our age range we socialized with. A big night was playing cards and when someone would bake a cake to go with the coffee and talk. When summer eventually arrived our entertainment expanded to include car rides and picnics, augmenting our card parties. Unfortunately the ship was in less than half the time.

"Credit for a junior Mackinaw crewman or his family was unheard of then. It was cash for everything. In those days the crew was paid in cash every two weeks on the first and the fifteenth of the month. Several times the ship came into the river and put a boat and crew over the side to deliver the cash to the waiting dependants. As soon as the boat was retrieved we turned around and went back out. I often wondered in those days whether the cash was to keep our loved ones from starving or to satisfy the local businessmen.

"One of the things we often did in those days when we had no money was to go down to the ship at night and watch movies on the mess deck. The berth deck would be full of babies sleeping while we enjoyed ourselves. The 'uniforms required' rule was relaxed then and we could come down in civvies if we were on liberty. Often Joana and Randy would come down when I had the duty providing I didn't have the 8 to 12 watch.

"During my time on the Mackinaw I ran the Ship's Exchange, which paid $25 per month. I saved everything I earned and eventually bought my wife a used washer and dryer. These would be needed in the future when more babies were expected to arrive. To that point in time she used a scrub board and the bathtub.

"Life in the Coast Guard can be hard at times, especially back in those years. In the early years of our marriage there was no support structure. You were expected to go and do what you were ordered to do and the families would have to make do as best as they could.

"Joana and I struggled when we were first married because of low pay, difficult duty, poor housing, a lack of medical benefits until the late 1950's, packing up and moving on short notice to start all over again in a different place away from family and friends, working on cutters whose time between overhaul periods became more and more frequent, with the loss of liberty that was the lot of a snipe until the work was completed.

"There is an old saying that 'If it doesn't kill you, it will strengthen you.' Joana and I would not trade those hungry years for anything."

"Our first house in Cheboygan was on Duncan Avenue, when it was a corduroy road," said Byron Trerice, on the ship in the early 1950's. "They laid down planks and covered it with soil and logs. When the trucks would go by you'd really feel the vibrations. The insulation was all sawdust, from the sawdust pile in town. At night you could hear the mice running around in the sawdust in the walls. But overall, it was all very delightful."

Fred Thurston and his family lived at 128 N. "C" St. in Cheboygan and always figured that his house was the closest to the ship.

"The other wives used to come over to our house when the ship would be due back because you could see the ship from our house," Thurston laughed. "We had two children graduate from Cheboygan High School and they enjoyed it very much. Our kids went to college up there too, so we've got some ties to Northern Michigan even though we eventually moved back to Ohio."

Holiday parties were also a big part of the crew's social life and were hosted at Thanksgiving, Christmas, New Year's Eve and Easter, subject to the ship's schedule allowing the crew to be in Cheboygan. Often the organizers would have door prizes and contests to boost crew morale.

Charles Dorian, a crewmember from 1959 through 1961, recalled that the ship was called out for Christmas Day of 1960 to break out an ore boat near Toledo. "We wound up spending the night at the Rouge River Dock in Detroit after we finished," Dorian recalled. "Other than that, holidays were

- - *Menu* - -

Chilled Tomato Juice
Shrimp Cocktail Cocktail Sauce Soda Crackers
Roast Young Tom Turkey with Oyster Dressing
Baked Virginia Ham Sweet Candied Potatoes
Whipped Mashed Potatoes Giblet Gravy
Buttered Fresh Frozen Peas
Buttered Fresh Frozen Green Beans
Cranberry Sauce
Spanish Stuffed Olives and Green Olives
Carrots and Celery Sticks
Lettuce with French Dressing
Parkerhouse Rolls Clover Leaf Rolls Butter
Fresh Milk Coffee
Minced Meat Pies Pumpkin Pies
Fruit Cakes
Delicious Apples Oranges Tokay Grapes Tangerines
Assorted Mixed Nuts Assorted Hard Candies
Apple Cider Cigars Cigarettes

1955 Christmas Menu

always pretty special. They usually had presents for the kids and a wonderful dinner aboard. The food was always very good."

Bob Brutcher reported to the Mackinaw from the 9th District Office in Cleveland and said this practice was continued into the 1980s, when he served in Cheboygan.

"They would cater the meal, and the Coast Guard Wives Club was active at the time and they used to do a lot of preparation and make a lot of things for the parties, especially Christmas," Brutcher recalled. "There would be presents for the kids and games. Sometimes we would do gag gifts for the adults."

One Christmas a sponsor stepped up to provide a prize package never to be forgotten. Larry Otto, the Mackinaw's executive officer, won a contest for the most children with nine. He and his wife were given a month's supply of free milk from the Inverness Dairy, worth $50 at the time.

For Thanksgiving Day dinner in 1949 the ship had a holiday meal for crew, family & friends.

"I invited a girl from town," Robert G. Swanson, Jr., remembered. "We all were sicker than dogs the next day from food poisoning. Did not make a good impression on that girl."

Sometimes the Mackinaw was out of town for holidays. This made things even tougher on families, who had precious little in the way of spending money in the first place.

Lynn Griffin, who provided a pay stub from

Ed Pyrzynski's retirement orders are read by Capt. George Winstein on the ship's fantail deck in 1966.

Pyrzynski's family came aboard for the Chief's retirement ceremony. With Ed are (from left) daughter Suzette, wife Dorothy, daughter Toni, and son Henry.

July 1968, when he first reported aboard the ship, showed that his E2 pay back then came to $65.70 after deductions. He also sent a work schedule for the week of Christmas 1969 that read as follows:

1. Full workday Tuesday.
2. Noon liberty on Christmas Eve 24 December.
3. Noon liberty on Friday 26 December even if we're in the Soo.
4. The ship's routine requires that the Mess Deck and the Library be secured at midnight. This means that all activity will cease from midnight until Reveille except for cooks and mess cooks.

There were a lot of long faces aboard the Mackinaw that year, Griffin said, as the crew also received orders to keep the Straits of Mackinac open until Jan. 15. There was no holiday routine between Christmas and New Year's either.

Griffin said that a sign was put up behind the Christmas tree on the Mess Deck, likely the most pathetic Christmas tree anyone had ever seen, saying:

TELL IT LIKE IT IS:

1. Bend over, it's Christmas.
2. Sandy Claws is dead.
3. Wait 'til you see the Easter Liberty!
4. What's the tree for?
5. Happy Ship, Happy Crew. Merry Christmas, Turn To!
6. Will the person with the dip stick for measuring morale please place same at bottom of Christmas tree.

Ruth Lorenz visited the ship for a dependent's cruise in 1957. The wife of Bill Lorenz, she was seven months pregnant then.

Captain Gordon Hall arrived for duty from Maryland in 1978 and brought with him his wife and four children, aged six through 12 years of age. The Halls were soccer enthusiasts, and there was no soccer in Cheboygan in 1978.

"My wife planted the seed that it would be great to have soccer in town," Hall recalled in 2005. "I started right in coaching Little League baseball, but she knew if the kids in Cheboygan got a taste of soccer, they'd go for it. We started with 60 kids in two age groups and it took off from there."

Hall said they scheduled soccer events around baseball so the kids could play both sports in the summer. He also credited Cheboygan High School Athletic Director George Pike – who coached football and went on to become the school's principal for many years – with giving the sport a boost at a time when some thought the evolution of soccer might cut into the numbers for high school football.

"The Rotary Club got involved and it became a real community effort," Hall said. "It was a good time and the crew and community were all working together, everybody helped get soccer started."

The Captain Gordon Hall Soccer Field complex was dedicated in 1984, right across the river from where the Mackinaw was docked.

Hall, like many family men stationed on the Mackinaw, chose to get everyone in his household involved in the community. He encouraged the same attitude with his crew.

Captain Gordon Hall Soccer Field, dedicated in 1984.

"My wife was in the Northland Players," he said. "Our kids walked to school and became involved in lots of school activities. The kids thoroughly enjoyed living in Cheboygan and even today we look back as a family and recall Cheboygan as being a great place to live, great duty for a Coast Guard family."

In 1980 the Coast Guard completed construction on a housing complex that rivaled any neighborhood in Cheboygan, producing a street lined with single family units and townhomes, spacious yards and ample recreational areas for families.

Jim and Shirley Honke became the first commanding officer's family to move into the new

Coast Guard Housing – Cheboygan, Mich., completed in 1980.

neighborhood after its commissioning, which looked far better than any government housing tract ever seen before.

To a Coast Guard family new in town, the housing situation produced instant friends with lots in common – everybody was used to the same way of life. The down side, some felt, was that it was too easy to stay within the ship's tight-knit community and miss out on meeting other people.

"I lived in both settings, in a house on Bailey Street that Captain Hall had lived in, and the Coast Guard housing," Honke said in 2005. "I really don't know which was best. There just wasn't an availability of housing in Cheboygan at that time in the early 1980s. It did isolate us from the community in a sense, and while it may have been good for the families to be together while we were gone it didn't allow us to get out in the community as I would have liked."

Honke wisely led many efforts to get his families involved in activities, including bowling, a softball team that played in a local league, and an enlisted wives club that met for social and community service purposes for several years.

When the housing development was completed, the wives club saw to it that the single street in the neighborhood was named Blackthorn Drive, in memory of 23 men who died on the U.S. Coast Guard cutter Blackthorn, WLB-391. The ship sank after colliding with an oil tanker Jan. 28, 1980 near Tampa, Fla.

"We enjoyed Cheboygan very much," said Brutcher, from Lakewood, Ohio. "It was a place where we didn't have to lock our doors, didn't have to worry about where our kids were. We almost stayed here for our retirement."

"People were really nice to us and it was easy for us to blend in with the community," Thurston added. "We enjoyed it very, very much."

From the beginning in 1944, the Mackinaw has had a tradition of community service in Cheboygan. The officers and crew are seen at church and school functions, involved in youth activities and taking on community service projects. For many years the ship sponsored a two-mile stretch of U.S. Highway 23 between Cheboygan and Mackinaw City – that commands a breathtaking view of the Straits of Mackinac's South Channel – for roadside clean-up.

Larry Otto, left, retired from the Coast Guard in 1971 and served a term as Cheboygan's mayor from 1972-1974, often hosting dignitaries aboard the Mackinaw as in 1974, the year of city's centennial celebration.

Christine Nickerson, Cmdr. Jonathan Nickerson and Adm. James Hull, the day Nickerson took command in 2000.

Anne Adams waves to the U.S. Coast Guard cutter Mackinaw with sons Teddy (age 1, in stroller), Roy (age 4) and George (age 6). The ship was returning to Cheboygan from the Coast Guard Festival in Grand Haven, Mich., reuniting Ensign George Adams with his family in 2005.

OFFICERS

C. R. MacLEAN, CAPT., COMMANDING OFFICER

J. E. D. HUDGENS, CDR., EXECUTIVE OFFICER

W. D. BALL, LT.
L. J. LARSON, LTJG
J. E. GRABB, LTJG
R. E. IDEN, LTJG
D. L. STIVENDER, LTJG
H. J. ROEHNER JR., ENS
W. G. GRAY, ENS
E. W. LEWIS, ENS
A. LANDRY, ENS
R. E. LARSON, ENS
R. J. HOLMSTROM, ENS
H. S. HAYMAN, CHPCLK, W-2
E. B. EATON, MACH, W-1

CHIEF PETTY OFFICERS

D. R. ABELLO, CSC
E. BERRISH, DCC (P)
L. M. BUCALOS, SDC
A. A. DORRIS, QMC
O. A. ERICKSON, ENC
T. H. GOLDEN, ENLC
J. A. KELLER, YNC
H. G. MEADORS, EMC (IC)
J. J. MURRAY, RMC (HF)
S. E. PHELPS, ENC
E. PYRZYNSKI, BMC (P)

1st CLASS PETTY OFFICERS

L. S. Bauchan, EN1
D. G. Bohr, EM1
F. J. Duch, EN1
A. R. Gulau, EN1
D. M. Holbrook, EN1
J. E. Holthouse, EM1
J. K. Jones, CS1
F. A. Munro, RM1
R. G. Skolweck, EN1
F. L. Stormer, EN1
W. J. Tomac, EN1

2nd CLASS PETTY OFFICERS

P. A. Hagerty, CS2
A. G. Hanustak, SK2
J. A. Howarth, RD2
D. E. Hutchings, EN2
L. Jorgensen, EN2
J. L. LaParl, RM2
R. D. Purdy, DC2
W. E. Quisenberry, EN2
W. D. Sandy, EN2
F. E. Shaw III, QM2
K. B. Stewart, ET2
R. J. Sullivan, BM2
W. R. Thomas, BM2
R. A. York, QM2

3rd CLASS PETTY OFFICERS

M. Asato, SD3
L. R. Carvalho, SD3
G. L. Crane, EN3
J. H. Friend, EN3
T. G. Gibson, EN3
R. E. Harju, EN3
C. R. Kahler, BM3
J. H. Kingeter, EM3
T. Kwiatkowski, EM3
W. E. Lorenz, RM3
G. T. Messick, QM3
C. F. Peterson, SK3
M. L. Riegel, GM3
A. A. Simko, DC3
C. M. Sutterley, CS3
D. E. Wright, YN3

NON-RATED MEN

R. Anderson, SN
P. M. Anloague, Jr., TA
F. Anthony, SN (HM)
R. J. Baldelli, SN
B. B. Bateman, SN
R. A. Biederstadt, SA
R. J. Blatnik, SA
M. W. Brott, SA
J. E. Campbell, SN
J. A. Carty, SN (SK)
D. S. Craft, SN (RD)
E. L. Cross, SN
C. S. de Castro, TA
D. A. Desimone, SN
A. P. Dizon, TN
G. E. Donachy, FN
R. E. Erfe, TN
R. J. Fitz, SA
A. M. Garcia, TN
G. A. Gault, SN
F. L. Harnisch, SN
F. R. Hedl, SN
P. J. Holland, SN
D. E. Irmen, FN
R. K. Jauch, SA
J. Kaptan, SA
C. E. Ketner, SA
L. R. Knopf, SN (RD)
J. N. Kolanko, SN
J. E. Koreen, SA
R. H. Lange, FN
V. H. Langevin, SN
J. P. Leddy, SN
J. E. Logan, SN
J. A. Long, SN (RD)
J. S. Matthews, SA
C. W. Mikolsky, SN
N. D. Mruk, SN
J. J. Murray Jr., SN
R. W. Newcomb, SN
L. B. Norton, FN (EN)
D. D. Pahner, SN
K. L. Patrick, SA
H. R. Prudencio, TN
J. Ramicone, SA
T. J. Rooney, FN
R. Russell, SA
R. A. Ryan, SN
E. B. Sales, TN
D. M. Sharits, FN
P. H. Sjoerdsma, SA
R. N. Steinert, SA
A. E. Suarez, TA
K. O. Sunstrom, SA
T. N. Villanueva, TN
W. J. Wall, SN
M. W. Weaver, SN
J. D. Winslow, FN
J. A. Yarling, SN

1955 crew roster.

Coasties and Townies

Over the years, hundreds of Mackinaw crewmembers married locally and became members of the community after they finished serving their tours of duty. Others married locally and moved on in the Coast Guard or to other careers. Some came to Cheboygan already married and decided to stay.

But the presence of more than 130 new men in town caused a stir in 1944, and the "newness" of incoming transfers continues to this day as something every single young man in Cheboygan is aware of. "Coasties" are competition.

The very idea that crewmembers from the Coast Guard came to town and began romancing the local girls caused trouble from time to time with the local guys.

Bill Baumann remembers when he first met his future wife, Faith, who was out for a night on the town with friends from her Mackinaw City hometown.

"All I did was talk to her and everybody was getting along fine," said Baumann, a sizeable fellow who retired to Mackinaw City in later years. "And the next thing you know here comes this big guy named J.C. Stilwell, sticking his nose up against mine and telling me to stay away from their girls. We must have stared each other down for a good minute or two, but we didn't fight. I thought we would when we went outside and he was waiting for me, but all we did was glare at each other.

"The next time I went out he was in the bar and came over and bought me a drink," Baumann grinned. "We've lived in Mackinaw City for years since and we still laugh about it."

Others did not emerge as friends after some confrontations, and occasionally a Mac crewman wound up on the wrong side of the law. The "all for one - one for all" mentality sometimes didn't mix well with a night of drinking downtown.

"The Club 27 was known back then as the Bucket of Blood – and for good reason," recalled Ellis Olson, a former schoolteacher and historian who served as the city's mayor. "Those fights used to spill out onto the streets and some of them were legendary, to say the least."

"I can't think of anything serious that ever happened during my years on the ship that didn't involve the bars," Lynn Griffin disclosed.

Ed Pyrzynski, who retired in Cheboygan, said the stories are mostly true. Many different bars were the scene of altercations between the men of the Mackinaw and local fellows.

"I was in a lot of those fights, it became almost a

regular thing back then," he said.

"When I came here in 1968, I met people around town who told me they would go out on Friday and Saturday nights looking for Coasties to beat up," Rene Zimmer said. "But we probably played a part in it too. Half of our crew was under the age of 20, most were away from home for the first time, and we partied hard and didn't care what happened. In later years the Coast Guard got a little tougher on alcohol and that swung the tide on behavior in town."

"I was from New York and figured I would be labeled," offered Frank Umbrino, "and I met people in town who took me in like a brother. I had no complaints about Cheboygan whatsoever."

Rod Cooper said he liked his assignment on the Mackinaw so much that he requested a second tour of duty.

"I met my wife the first time, but for two years she wouldn't go out with me," Cooper recalled. "When I came back after a year at Belle Isle, she finally went out with me and we got married. I was glad because until then I used to get into a fight about every night, it seemed."

Apparently Cheboygan wasn't the only town where Mackinaw crewmen found rivalries with local "townies." Cooper said DeTour, Mich., will never forget a night in the late 1960s when the Mackinaw pulled into port and granted a "Cinderella liberty" after three weeks of breaking ice.

"There were only two bars in town, as I recall," Cooper said. "There were about 60 to 80 of us who went into one of them for the evening. The townies didn't like us, and we had a lot of problems that night. Supposedly we aren't allowed in DeTour to this day, I don't know."

However, the Mackinaw's crew usually became known as good citizens. Those who didn't pay rent, didn't pay bills or became disruptive got into the

Richard and Donna Hess were married in 1948 after meeting in Cheboygan. A wedding date could change if the ship was called out on a mission.

same troubles that anyone else would. Some landlords preferred to deal almost exclusively with Coast Guard tenants, due to their reliable reputations, the common term-lengths of their hitch, and the knowledge that if a problem arose the landlord could always go to the captain of the ship.

Over time, attitudes changed as more and more Coast Guardsmen married and remained in the community to become solid citizens.

Many crewman rented onshore to avoid living aboard the ship.

"I had several people in town ask my wife to introduce their daughters to Coast Guardsmen," Pyrzynski added. "So it turned around, eventually."

Sometimes it was the landlords who were known for their quirks -- and their kindnesses.

"I've met several guys at the various reunions who all lived in the same house on Mackinaw Avenue," stated Byron Trerice. "The lady who owned that house was called Ma Burrows, and used to only put 25-watt lightbulbs in the fixtures. Then she would go outside and watch the meter running and she claimed she could tell if you had switched to a higher-wattage bulb. We used to bring better bulbs from the ship, but when we were out of town she'd come up to the room and change the bulbs. It was the times, and we didn't say anything about it."

During the 1974 Reunion and groundbreaking for the new moorings facility, the Reunion Committee honored Mabel McNellis, who rented rooms at her property at 217 N. E St. to Coast Guardsmen for years. She was introduced as the housing "mother" to many Mackinaw sailors over the years, some at $2 per week.

Over the years, some Mackinaw crewmen had their share of entanglements with the law, as is bound to happen in any society. The incidents, however, were few and far between. When something did happen it tended to stick out because it wasn't the norm for the crew's behavior as a whole.

In the early years, a Mac crewman drove a car for some time that had been stolen from another city. He took it everywhere, including a trip south through Ohio, Kentucky and Tennessee. On one last trip to Chicago, his irate shipmates became aware of the car's origin and suggested he turn himself in. The car was left on the highway near Grand Rapids, where it was originally stolen, and the thief got everybody to wipe down the door handles and the interior of the car to remove any fingerprints. The gang then hitchhiked the rest of the way.

However, the police found an identifying item under the seat of the car and showed up at the ship to question the sailor. He denied it at first, but when presented with the evidence knew they had him and admitted his theft.

The ship's storage facility at the Millard D. Olds Moorings was pillaged more than once, and police cooperated with the ship's command when the culprit was found to be a crewmember. A refrigerator, a lawnmower, a washing machine and other sundry supply items went missing at one time or another, resulting in a trip to the Brig, the Cheboygan County Jail, or both for the perpetrator.

Sometimes unpaid rent was the tip of the iceberg. Once, a complaint from a landlord revealed upon inspection a stolen motorcycle and other items. Civilians who were involved had quickly left town. The sailor was apprehended and paid the price.

A particularly embarrassing episode involved a new crewman who was lodged temporarily at a downtown motel and began playing around with a BB gun, taking potshots at a bus parked outside. It turned out the bus was transporting an entourage that included the governor of the state of Michigan, also staying at the inn. The governor's security force caught the sailor. His shipmates wondered why he needed a BB gun for duty in Cheboygan.

Once an irate crewmember chain-sawed the Mackinaw's sign at the dock as revenge for being found out in an on-board romance (and various other problems) and banished from the ship. He also damaged an officer's truck, parked at the dock. Three spent rounds from a .357 revolver were found in the vehicle.

These incidents are not recalled to bring shame to the crews of the Mackinaw. They were all told by those who served on the ship and are simply part of the vessel's history. Mac sailors were human beings, like everyone else, and sometimes made mistakes.

Although the Coast Guard stood for marine law enforcement and was a local representative of a branch of military service, the Mackinaw found itself in hot water in the early 1970s.

A campaign was underway to clean up the Great Lakes at the time, when plenty of attention was being given to filthy water conditions at many port cities including Cleveland, Detroit, Buffalo and Milwaukee. State and federal laws were being changed to regulate pollutants being discharged into America's waters and air supply.

The Mackinaw was still "old school" at that time, and dumped overboard plenty of sewage as all ships had been permitted to do in years past. The Mac, however, was accused of doing this right into the Cheboygan River. City fathers noted that there was no apparent hook-up to a sanitary sewer or septic system.

Cheboygan County Prosecuting Attorney Jerry Sumpter threatened Capt. Lilbourn Pharris, Jr., with a trip to the County Jail unless the ship obeyed local sewage laws at the dock on the river's turning basin. Pharris cited federal designation as a Coast Guard ship and told Sumpter he had no jurisdiction over the vessel or his command.

Eventually the ship was properly connected to a pump-out tank at the dock and tensions eased. Today's Coast Guard regulations would make such an incident quite unlikely to occur, but in those times things were different. Many large corporations had to adjust their way of treating and discharging waste materials.

In the later years of the ship's service reports surfaced that some Mackinaw crewmen of color had experienced attitudes and prejudices that were, unfortunately, typical of small towns in the north that had little, if any, diversity in their racial make-up. It wasn't as though there were public acts of hatred toward Coasties with a different color of skin – but there was certainly ignorance.

In 2003 the city of Cheboygan, with the specific experiences of some Mackinaw crewmembers in mind, established a Human Relations Commission to study and attempt to solve racial issues within the community. The Commission found that the scope needed to be widened beyond the Coast Guard's presence in town, and an education process began that was beneficial community-wide. It extended through the school system and beyond to many community organizations.

In 1963 the city of Cheboygan re-named North B Street, the street leading to the Mackinaw's dock, as "Coast Guard Drive." Jim Muschell was in his first term as mayor at the time, and Capt. Ben Chiswell was the ship's skipper. Bob Henderson was a key force in getting the name change done as the military affairs representative for the Cheboygan Chamber of Commerce.

Lt. j.g. Al Landry led a marching unit from the Mackinaw in the 1956 4th of July Parade in downtown Cheboygan. This photo was taken at the corner of North Huron Street and Backus Street near the Opera House, looking east on Backus towards Main Street where the parade had turned. The Steffins Block is visible in the background, and the buildings on the left side of Backus were torn down long ago.

Mascots of the Mackinaw

It is very unlikely, to say the least, that you would see a dog on board a vessel in today's Coast Guard. Times have changed from the days when a pet was kept on a ship for companionship or good luck.

Actually, animals have served as mascots on board Coast Guard vessels since the early days of the Revenue Cutter Service. The practice of keeping pets may have started when cats were brought on board to combat the rat population. But for years, pets have helped keep the crew's morale high during their many lonely days at sea.

During the first half of this century, nearly every ship had at least one mascot and some had menageries that were the envy of a small zoo. According to Coast Guard historian Florence Kern, Captain Mike Healy – commanding officer of the Revenue Cutter Bear – kept his parrot on board for company.

Dogs have been the most common mascots through the years, and one of the most famous was Sinbad, who served on board the cutter Campbell during World War II. He came on board the ship in 1937 when the Campbell made a port call in Portugal. Sinbad remained on the ship throughout the war. A "salty sea dog" all the way, Sinbad stood watches, ate his meals and slept with the crew. Sinbad was as much a part of the Campbell as his two-legged shipmates. His contributions to that ship were incalcula-

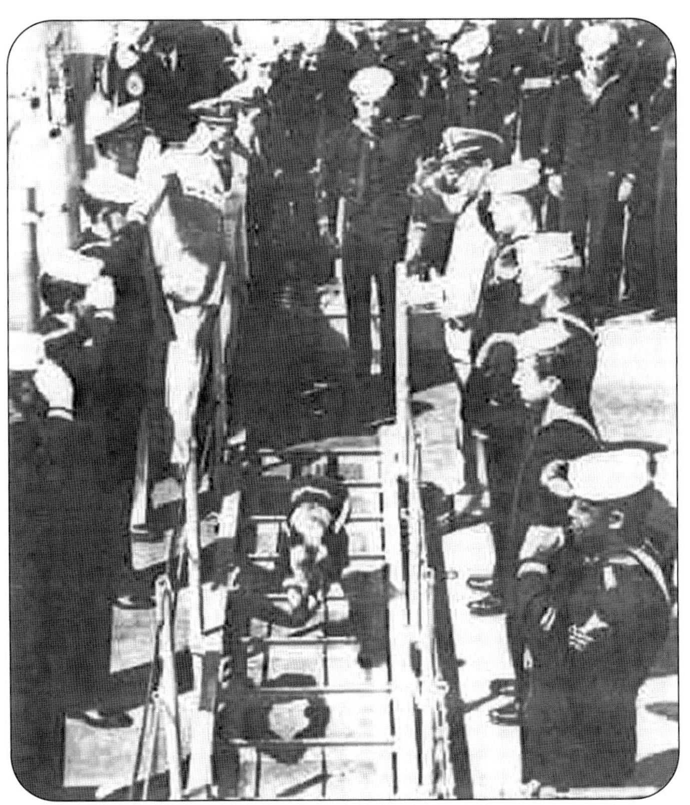

When "Max" retired from the U.S. Coast Guard cutter Klamath, the officers awarded him a rank of Petty Officer 1st Class. He was piped ashore as the crew saluted the long-time mascot.

ble in terms of the morale boost he provided. To his shipmates he was their talisman, their good luck charm that brought them through battles with submarines, storms, and the icy North Atlantic.

The Mackinaw, too, had several legendary canine residents on board during the early years of the vessel's history. They each had their own personalities, and different crewmembers recall different dogs with varying degrees of affection – or dislike.

Jack Eckert served aboard the Mackinaw and for many years contributed to a fabulous Web site entitled "Jack's Joint" at www.jacksjoint.com. Jack contributed a story from Steve Paddock from the years 1953-1954 when the crew first had a mascot aboard. It was a dog that had been picked up as a puppy off an ice floe. He was, of course named "Mac."

"We were near Lansing Shoals in Lake Michigan," said David Kaplan. "They spotted a brown collie on an ice floe. We used a personnel boom to swing me out and I climbed down the Jacob's ladder and I put a sling on the dog and brought it aboard. It was a nice dog, we spoiled the heck out of it."

When a new crewman reported aboard the dog would stay close to him and would sleep under the new crewman's rack. Mac then treated that person like the rest of the crew. Mac recognized that people in a Coast Guard uniform were OK even if they were coming aboard for the first time.

During Paddock's tour of duty aboard the Mackinaw the ship got a new executive officer.

"He arrived the weekend before his assignment date and decided to visit the Mackinaw in civilian clothes, he came alone," Paddock recalled. "As soon as he arrived at the top of the gangway our faithful Mac attacked this stranger and actually bit him on the buttocks. Well, needless to say, after that no matter what uniform the exec put on Mac would have none of it. As far as Mac was concerned that man was an intruder. Whenever the crew would muster on the quarterdeck and the exec was present the dog had to be restrained. It became a matter of great amusement for the crew and an embarrassment for the exec and some of the other officers.

"As it turned out Mac and the exec simply could not be crewmates on the same vessel and Mac was given to one of the crewmembers that lived on a farm near Cheboygan. The crew was quite upset about this because we considered Mac ours. In the end it all came to a happy conclusion when we were allowed to get another mascot. It was trained not to eat executive officers!"

Blackie was another dog who served aboard the ship and came aboard at Cheboygan with a

Blackie on deck, helping to clear ice in 1948.

crewmember that soon departed and didn't take the dog along. From that day forward, Blackie was cared for by Willard Teare more than anyone else.

Knowing that a shipmate likes something can sometimes set him up for a practical joke or a prank, and Teare's daughter Shirley remembered an occasion when the Mackinaw was in Grand Haven for the Coast Guard Festival. Someone stole Blackie and thought it would be funny to hide her on board another cutter docked there. Several members of the Mackinaw crew joined Teare and conducted a search for the dog, finally "stealing" her back.

"My father loved pets," Shirley recalled, "so he adopted her. Wherever he went on the ship, so went Blackie. When my father retired from the Mackinaw, he brought Blackie home with him and she became my dog while I was growing up."

Eckert also related a few stories that he had about

Blackie in her retirement years with Willard Teare, her Coast Guard owner.

Maggie, a dog of mixed popularity who was on board the Mackinaw in the mid-to-late 1950s. He contributed several from Floyd Stormer that probably take the prize.

While most mascots living on Coast Guard ships were well loved and cared for, "Maggie of the Mackinaw" got to the point where she didn't like anyone. She was at least part German shepherd, but some would say "part" pit bull. The men of the Mackinaw returned her animosity in triplicate. The biggest problem was that one could not get rid of the other. Yet dogs don't usually get mean unless they are trained to be mean.

Members of the crew say that a first-class engineman – who later made chief – continuously tormented the dog and taught her disgusting actions. She would go around and hump the leg of someone sitting down. She did the same thing to a foul weather jacket tee-pee'd for her trick. The tormentor continually cursed her and spread tales about her and her alleged escapades. He evoked many a laugh from his own crowd for his antics and thought himself quite the hero.

It is regrettable that during that era there was a greater division than normal between the officers and the crew. Obviously some of the officers were vaguely aware of what was going on and took no action — just as none of the tormentor's peers and juniors failed to step in. According to Stormer, the man was an intimidator and in those days juniors did not question the actions of seniors to their faces.

In any event, Maggie insisted the Mackinaw was her home and refused to leave. Once she was left behind in Cleveland — accidentally, of course. However, the captain was certainly not going to go back to get her, even when it was apparent that the dog had jumped off the pier into the water. Maggie was seen swimming past the Cleveland Lights, trying to catch "her" ship. When the men at the light radioed the Mackinaw and told them their dog had already swum six miles and was still going strong, the ship was finally turned around to retrieve her.

Maggie thanked the crew by becoming more cantankerous than ever.

Tom Dellabeck holds "Maggie" of the Mackinaw, perhaps the ship's most loved – and hated – mascot. Steve Paddock is holding the broom for clean-up.

There was a time when members of the crew tried to "dump" Maggie by putting her off at DeTour Island while passing through the St. Mary's River channel. It was felt she was gone forever this time.

However another cutter, the Tamarack, spotted Maggie and knew she was the same dog from the Mackinaw. The Tamarack's crew didn't realize she wasn't exactly the most popular mascot in the Coast Guard at the time. They made a special trip to find the "Great White Mother" which was breaking ice in Whitefish Bay, to return Maggie.

If the men of the Tamarack expected thanks, they must still be waiting. To quote one man who was a member of the Mackinaw crew at the time, "that damn dog had more lives than ten cats."

Maggie also distinguished herself through one particularly noteworthy performance while the Mackinaw was visiting the shipyard in Manitowoc, Wis.

According to Eckert, Ken Patrick served on board the Mackinaw from October 1955 until May of 1957.

The way Patrick tells it, "That time in the Coast Guard changed my life. It was a time to grow up. Often people pass through our lives and we fail to recognize the influence they have. Many played a role in helping me to determine what direction my life would take, including a dog named Maggie.

"I guess I never saw the bad temper of Maggie that was portrayed by some people, maybe since I was in the commissary department and often fed her left-overs," Patrick said. "We all loved that dog when she was younger.

"I remember one time in 1957 when the ship was in dry dock in Manitowoc. It was my last trip on the Mackinaw before being transferred to the U.S. Coast Guard cutter Coos Bay in Portland, Maine. On the first afternoon after arriving in Manitowoc, three of us decided to head to a pub and we took Maggie with us. We traveled to town in a cab. She was a hit, as we were also. If you remember, everyone loved the Coast Guard and they adopted our mascot, too. She was given leftover beer by patrons and seemed to be having a great time.

"After several hours of partying, we met some girls. We were ready to continue our tour of town but now with female escorts. Sadly, there was no room for Maggie. We all tried to convince each other it was their duty to take Maggie back to the ship but to no avail. We finally came to the conclusion that we couldn't leave without our new friends. We had to find a new solution for Maggie.

"We finally did. Maggie, was put in the back seat of a cab by herself, a little tipsy perhaps," Patrick concluded.

The cabbie probably thought he'd seen it all until he was told to drive the dog to the shipyard dock, open the door and let her out.

"We paid the cabbie and gave him a nice tip, and off went Maggie back to the ship. We often wondered what other drivers thought pulling up to a traffic light, seeing a dog in the back seat of a cab. Nothing but first class for our mascot, Maggie!"

For years afterward, the crew of the Mackinaw told the story of the night Maggie showed up in a cab, jumped out the door and trotted across the yard and up the brow to the ship!

Al Landry contributed a follow-up story to Eckert about Maggie, telling of her eventual end. He said that the dog, well on in years, eventually became unable to walk without pain and required more care than the crew could give under the best of circumstances. He finally took the mascot to a veterinary office in Milwaukee, where Maggie was peacefully granted an end to her pain and relieved of duty.

And now you know the rest of the story about "Maggie of the Mackinaw," the once-loved ship's mascot that nearly everyone who served in that era has a story about.

Maggie on deck with one of her pups. On New Years Day, WCBY Radio held a contest to determine the first birth of 1957. The Mackinaw's crew thought Maggie's litter, born early that morning, should qualify and entered her in the contest. The ship's newsletter later reported that since Maggie could not identify the father, she was disqualified. The dog was noted for rushing at landing helicopters and snapping at the exhaust pipes. She rarely had whiskers...

USCGC MACKINAW 11 WAGB-83

All In A Day's Work

There were certain benefits to being assigned to the Mackinaw, for the crew never knew from one day to the next what sort of excitement might occur on the ship.

Jack Eckert said that James Waesche told a story of the ship's crew being offered the opportunity for some "Northwoods R&R." The story is repeated here with a courtesy to Esther Stormer, who published it in 1985.

"I was assigned to the Mackinaw in 1945," Waesche began. "The ship left Lorain, Ohio to go on up to Detroit, Lake St. Clair, Lake Huron, and into Cheboygan. This was in late September.

"In early October the ship took off for Lake Superior and had to pass a place between Upper Michigan and Drummond Island known as the De Tour Passage. The skipper, Cmdr. Roland, decided it would be nice for the crew to have a little R&R. This consisted of allowing any of the crew who wished to go deer hunting the next day on Drummond Island the chance to do so.

"So the next morning at 0500, a cold and windy 0500, we loaded to motor lifeboats," Waesche continued. "The water was almost coming over the gunnels. I thought we would never make it to the island. We split up into pairs and my partner was the ship's doctor. I was not sure whether that was a good or bad omen.

This Air Force helicopter landed on the Mackinaw's deck in Buffalo in 1950.

"The rule was you couldn't fire at a deer until sunrise. At sunrise all hell broke loose — there were rifles going off all around us. It was like being in no man's land. Most of these guys had never been deer hunting. The Doc and I wandered around, seeing nothing, colder than a well digger's clavicle. It was windy, really brutal. I kept thinking about the guys who stayed aboard that must have known something that I didn't.

"Finally, a rabbit jumped out and the Doc fired and hit him with a 30.06 dum-dum bullet. All that was left was an ear and the tail. At 1600 it was time to head back to the ship. Nobody even saw a deer, but a lot of shooting went on.

"As we neared the ship we spied something hanging from one of the cranes," Waesche said. "It was a large buck — completely skinned. While we were running around that island conducting our Chinese fire drill, a doe and a buck came swimming by going to another island. A boat was lowered and the buck was hit in the head, taken aboard and skinned. You can guess who ate venison and who ate crow."

During the mid-1950s, the Mackinaw's icebreaking capabilities were called upon to keep large pile-ups of ice from forming around the caissons being set into place to form the foundations of the Mackinac Bridge.

Don Wright took this picture of an icy Mackinac Bridge caisson from the Mackinaw's deck in 1955.

Michigan Gov. Kim Sigler toured the ship with his wife at Mackinac Island.

"We would go out there into the Straits when the ice got bad and clear circles around the bridge caissons," recalled Don Wright, who served as a yeoman during his years on the Mackinaw. "Rather than make the trip back and forth to Cheboygan, we would anchor out off Mackinaw City for several days at a time."

The ferry services were still running then, so the icebreaker couldn't utilize the State Dock at Mackinaw City as it did in later years after the bridge was completed.

"The icebreaking work would go from dawn to dusk, then we'd wait at anchor until morning," Wright said. "One time, a link broke on our anchor chain out there and we lost an anchor."

The Mackinaw's work allowed the bridge foundations to settle into place without the pressure of ice windrows piling up against them.

Larry Otto served from being an ensign all the way up to the ship's executive officer, and figured his time on the Mackinaw was all behind him when he scheduled his retirement for the spring of 1971.

"I'd planned my retirement out some time before, and assumed I would go off the ship while it was in Cheboygan," Otto said. "It's a nice ceremony and I was looking forward to it.

"Well, when the time came we wound up down in Lake Erie breaking ice. We were way out in the lake west of Buffalo, and the captain, Lilbourn Pharris, asked me if I wanted to extend my service for awhile and retire when we got back. My replacement had already been on board the ship for a week, so I wasn't in the mood to stay even one day longer.

"I told the captain, 'get me out of here,' but I didn't know how he could arrange it with a boat because the ice had everybody pretty socked in. We were out there helping some freighter get unstuck and finally it was time for my retirement ceremony. The captain came out and conducted the proceedings and it was all very nice.

"Then when we finished here came a helicopter, and it dropped right in and picked me up and away I went," Otto laughed.

"I piped him ashore that day, and traditionally you go ashore across the brow from the ship to the dock," said Fred Thurston. "I had the men move the brow across the deck so Mr. Otto crossed it right on the ship and then went up in the chopper."

Larry Otto's retirement — by H-52 Helicopter — in 1971

Joe Etienne recalled the night the ship was downbound across Lake Superior for Sault St. Marie from Duluth, running hard 20 miles from shore in a blinding snowstorm in the middle of the night.

"All of sudden there was a huge crash that stunned everyone on the bridge," Etienne explained. "Captain Roland came rushing out of his wardroom and said, 'Joe you must have hit a fishing tug or something, stop and check it out.' Well, the noise and the impact we felt on the bridge was so bad, we could have hit just about anything but I didn't think we had.

"I told the captain I had just checked my radar and we were clear," Etienne continued. "All I could see were snowflakes. I got the deck crew to man the searchlights but there was nothing around us, no ship anywhere in all that snow. Just about then a guy came up to the bridge and told us the ship's bell had worked loose and dropped down on the metal deck. That's what made all the noise, and where we were and where Captain Roland had been sleeping it sounded like a bomb had gone off. The vibrations of the ship had worked loose the nut holding the bell in place and it just dropped right down on the deck. The shipyard didn't put a pin or a key in there to keep that from happening.

"Captain Roland said, 'I'm going back to bed' and that was the end of that!" Etienne laughed.

In 1948 Roy Hillegen, known as "Mr. South Africa," came to town and impressed the crew with his physical strength and agility, walking on his hands across the large crane beam on the afterdeck and by lifting a large set of barbells.

The ship's bell never worked loose again once a sealed locking nut was installed.

A broken propeller blade sidelined the Mackinaw into dry dock for repairs in February 1977.

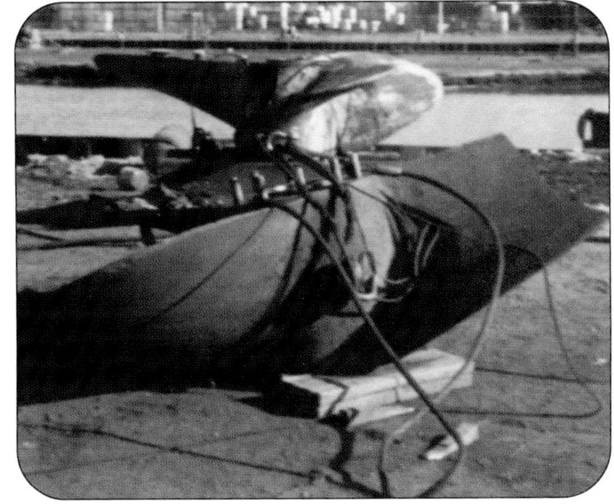

The propeller blade broke when the Mackinaw backed down its power in the ice during operations in the St. Mary's River, near Point aux Frenes. The repair was completed in drydock, with the broken prop pictured (above) and on the dock (right) with the replacement ready for installation right behind it.

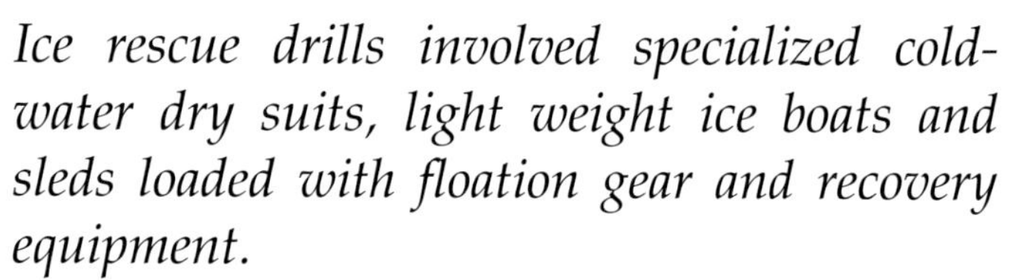

Ice rescue drills involved specialized cold-water dry suits, light weight ice boats and sleds loaded with floation gear and recovery equipment.

The ship's galley prepared the best chow in the Coast Guard for more than 60 years.

Food service on the Mackinaw was always good, depending on who you talked to. After a particularly rough day of icebreaking or any duty where the weather was extremely cold, the Mac's cooks could really hit the spot with the food they served. This was due to the size of the galley and equipment carried on board the ship compared to smaller cutters, and some of the best cooks in the Coast Guard. Of course, there were always complainers, and admittedly some cooks were better than others. And some crews just complained more than others!

Officer's mess.

The mess deck, full of Coasties enjoying another delicious meal.

Right - The original engine controls occupy a prominent position to the right of the wheel. The massive brass handles are linked to the ship's three electric propulsion motors: port aft, bow, and starboard aft.

The mouse and keyboard of the ship's modern navigational computer system are more recent additions to this area. The captain's chair, in which by tradition only the ship's commanding officer may sit, provides the C.O. with a comfortable seat during long watches on the bridge.

Two of the Mac's six 2,000 horsepower Fairbanks-Morse engines

The starboard stern propeller shaft.

Foward Shaft Alley, driving force of the Mac's novel bow propeller.

The commanding officer's cabin provided comfortable furniture for meetings, relaxation, reading or playing cards. Contrast these 2003 photos with the one below at left.

The CO's cabin, prior to the 1982 re-fit.

The fleet commodore's stateroom.

The ship appears frozen into place at the pier in Sault Ste. Marie, Mich.

A bow-shot taken in Lake Superior's Whitefish Bay from on the ice.

Easing the Big Mac alongside the pier at Detroit. Visible on shore are the Joe Louis Arena, Cobo Hall, and the Renaissance Center.

Ship meets ice in this photo of the Mac's bow doing its job.

Search and Rescue

The Mackinaw was not a ship that would handle routine search and rescue duties unless she was already at sea. It would be pointless and inefficient to go through the entire procedure of getting the giant icebreaker ready for duty when a rigid-hulled inflatable boat or even a smaller cutter could arrive on scene in the Straits of Mackinac, Lake Michigan or Lake Huron in a fraction of the time.

Coast Guard stations on Mackinac Island and later in St. Ignace usually handled this sort of duty on a regular basis during most of the years the Mackinaw was in operation. Today, the Coast Guard Air Station at Traverse City can respond to search and rescue calls much faster and will respond if there is too much travel time involved from other Northern Michigan facilities.

For the Mackinaw to be involved in search and rescue the ship needed to already be underway, perhaps escorting the fleet for the Chicago-to-Mackinac race or on the way to a festival in a Great Lakes port.

The ship did keep a small motorboat ready in the davits for a quick launch into the river where she was moored, but in general when the Mac answered a search and rescue call from Cheboygan, you knew it was a big deal.

The first really big search and rescue mission for the Mackinaw came during the early morning hours of June 24, 1950 when a Northwest Airlines DC-4 exploded and crashed into Lake Michigan 20 miles north of Benton Harbor, Mich. All 58 people aboard died.

Although eyewitness accounts reported a bright flash over the lake and an explosion, by 5:30 a.m. that Saturday morning Northwest Orient Airlines Flight 2501 – heading from New York to Minneapolis – was

The Mackinaw was called to the crash scene of a Northwest Airlines DC-4 that crashed into Lake Michigan in 1950.

presumed lost as the fuel supply would have been exhausted by that time. At daybreak, the search and rescue teams began an intense search on the fog-covered lake.

The U.S. Navy, U.S. Coast Guard and state police forces from Illinois, Michigan, Wisconsin and Indiana all joined the search. Some 13 hours later – at 6:30 p.m. Saturday evening — the U.S. Coast Guard cutter Woodbine found an oil slick, aircraft debris, and an airline logbook floating in Lake Michigan many miles from shore. At 5:30 a.m. on Sunday, June 25, sonar work by the U.S.S. Daniel Joy near the oil slick revealed several strong sonar targets.

In addition to the Mackinaw, the Coast Guard vessels Woodbine, Hollyhock and Frederick Lee focused on the recovery of floating debris. This included a fuel tank float, seat cushions, clothing, blankets, luggage, cabin lining and, tragically, body parts. At the time, authorities wanted to determine whether the plane suffered a mid-air explosion, or whether it struck the water intact. These small pieces would be the only clues they had.

Small bits of debris floated endlessly over the surface of the fogbound lake. The airplane, along with 58 men, women and children had disappeared, leaving few clues as to what had occurred 3,500 feet in the air. The loss of Flight 2501 represented the worst commercial aviation disaster to that time. For two weeks after the disappearance of the plane, human remains, clothing, personal effects and debris washed ashore all along Allegan County's coastline.

Initial reports suggested the plane exploded in mid-air, with debris falling into the lake between Glenn and South Haven, Mich. Officials began discovering debris and body parts Saturday and Sunday over a four-mile area about 12 miles northwest of Benton Harbor.

Captain Carl G. Bowman, skipper of the Mackinaw at that time, told the United Press bureau at Detroit by radiotelephone that "Tiny pieces keep floating to the surface all through the area." He said his men found hands, ears, a seat armrest and fragments of upholstery.

One week later, portions of the bodies of two women were discovered – one about two miles north of South Haven and the other about seven miles north, at Glenn, Mich.

"I was on the Woodbine when they found the pieces coming up," said Larry Otto, at the time a 23-year-old junior officer with just two days on the job serving on the buoy tender. His first memory of life in the Coast Guard was heading out on the ship to look for survivors of the tragedy. Otto was later an executive officer on the Mackinaw.

"It was all pieces of airline seats or small pieces of insulation attached to pieces of the airplane," Otto recalled. He soon realized there weren't any survivors to be found. "We found a baby's heart on a blanket attached to a seat; we found a lung without the rib cage on another part."

"It was horrible," said David Kaplan, a boatswain's mate on the Mackinaw. "We picked up 32 buckets of pieces of remains and stored them in the walk-in cooler until the medical examiner came and took them. We found a lady's purse that looked intact until we opened it. The glass in the small make-up mirror was pulverized."

Another time the Mac was in Lake Michigan for the Chicago to Mackinac race and the crew came upon a body floating in the lake, the result of a small plane crash some weeks before.

"It's unpleasant work, but that's what the Coast Guard does," said Rod Cooper, a boatswain's mate in the late 1960's. "The poor man had been in the water about three weeks and we were having trouble getting him in the litter. He kept floating away. One of

A Piper Bear-Cub was pulled from Lake Michigan near Harbor Springs by the Mackinaw's crane in the late 1940s.

the guys said, 'I'll get him,' and pulled out a boat hook. I shouted 'no' but it was too late. Every one of us on the small boat got sick over the side."

Byron Trerice recalled the mission to rescue the freighter Maryland near Marquette, a mission to tow it onto a sandbar to save it from sinking.

"The Maryland had about 16 feet of water in its bilges when we got to it," said Trerice, who served on the Mac in 1953 through 1954. "We got it onto the sandbar, then had to go after a barge that had separated from the tug Favorite. The experience was rather remarkable in that we were recording winds at the top speed we could measure on the bridge and had waves in excess of 30 feet. The ship was listing in excess of 35 degrees, literally on its side.

"At the time we had regular lifeboats in davits, and those boats were ripped from the ship and the cleats were torn right out of the metal deck," Trerice continued. "We also had a gas tank welded to the deck for the occasional time that we would have a helicopter land on board. The waves ripped that gas tank off the deck and it was lost at sea. I think there were only three people that I was aware of on the ship who didn't get sick that night. I was up for 36 hours and somehow that feeling passed and I never felt so good in my life."

Charles Dorian said that the Mac was pressed into duty for a Greek freighter that missed the turn at Gray's Reef Passage and ran aground at Beaver Island.

"We went down there the next day and took the crew off the ship," he said. "We rescued them, and the captain too. His wife was there too and we brought them all to the Mackinaw. Another time we were called to look for three teenagers in the water near Bois Blanc Island, and when we approached from the east we saw what appeared to be a black five-gallon can in the water. It turned out to be the head of a young lady swimming very slowly. She said the other two had gone ahead towards shore. I'm sure we saved her life that day."

Of course, there were also false alarms. Jack Eckert recalled one that came at a very inopportune time.

"As Thanksgiving neared, all dependants were invited to the ship for dinner," Eckert recalled. "Nobody bought food for home. The day before Thanksgiving we were bowling in our regular Mackinaw league at the local bowling alley when a member of the duty section in undress blues came in and went directly to the Captain, who then asked for our attention.

The car ferry Vacationland ran aground in the Straits of Mackinac due to large heavy ice floes and high winds in the spring of 1956. The Mac snapped a three-inch towing cable and damaged the winch clutch trying to free the vessel. The cutters Acacia, Kaw and Arundel helped to de-fuel the Vacationland and the Mackinaw removed cars from the ship with her giant cranes to lighten the draft. The three-inch cable was re-tied in a figure-eight arrangement around the towing bits and the ship was pulled free. The Vacationland later sank in a Pacific Ocean storm, swamped by huge waves that broke open her giant deck doors while being towed to China for scrap.

"The District, he told us, was sending us on a search and rescue mission a couple of hundred miles away and we were sailing in three hours. Our dependants didn't get Thanksgiving dinner, and ours didn't taste too good either. We didn't even find what we had been sent out for."

Some missions occurred spontaneously, while the Mackinaw was already at sea. Ed Pyrzynski recalled one with a happy ending in Lake Michigan.

"We were breaking ice in Little Bay De Noc, near Escanaba, in 1948," Pyrzynski remembered. "Along came this good-sized ice floe with a car on it. The people were waving their arms, just floating along. They were ice fishing and the ice broke apart and they were in trouble."

The Mackinaw would be the only answer to a predicament like that, with exactly the right equipment for the job.

"We brought the ship right up to them, lowered the crane and picked up the group and their car," Pyrzynski laughed. "They were glad to see us."

The Mackinaw missed the 1958 disaster that came when the Carl D. Bradley sank in northern Lake Michigan.

"We were being overhauled then," Capt. Joseph Howe said some years later.

The Mac was quick to respond on Nov. 30, 1960 when the SS Francisco Morazan ran aground on South Manitou Island near where the Bradley sank.

"We were there the whole time," Howe recalled, "and the men took the crew off in small boats during a howling gale."

On the foggy morning of May 7, 1965, the 588-foot freighter Cedarville sank after being struck near the Mackinac Bridge by a Norwegian vessel, the Topdalsfjord. Many police agencies and the Coast Guard responded while a German vessel, the Weissenberg, stood by at the wreck site.

Jerry Pond was a 17 year-old student at Cheboygan Catholic High School, then located on Dresser Street in the downtown area, when he was summoned from class by Alta Riggs, wife of Cheboygan Daily Tribune publisher Myrt Riggs.

"I was taking pictures for them after school in those days," recalled Pond, who remained at the Tribune and in later years became a pressman. "Mrs. Riggs sent word to get me out of class and an employee, Chris McGurn, brought a camera and told me

The Mackinaw took survivors and casualties of the Cedarville from the German Ship Weissenberg.

Two survivors of the Cedarville, Ralph Przybyla and Stan Mulka, after being transferred to the Mackinaw.

Cedarville casualties were covered with American flags onboard the Mackinaw.

there had been a shipwreck. She drove me to the dock and I got on board the Mackinaw just as they were bringing in the brow and went out there with them.

"I had never seen anything like that, dead bodies and all. It was very foggy and you could barely see the other side of the Cheboygan River as we were leaving town. I just took pictures of what was happening while they transferred the men from the Weissenberg over to the Mackinaw and picked up some from the water. We carried the survivors and the deceased to the dock in Mackinaw City."

One officer with a Mackinaw connection, Lt. Jack Rittichier, became the only member of the U.S. Coast Guard to be classified as "missing in action" in Vietnam.

Originally in the Air Force, Rittichier accepted a commission in the Coast Guard Reserve as a lieutenant, junior grade, following his discharge and was eventually promoted to lieutenant in the regular Coast Guard. Rittichier and his unit were awarded the Coast Guard Unit Commendation for their rescue work during Hurricane Betsy.

While assigned to Air Station Detroit, based out of Selfridge Air Force Base, the Coast Guard awarded Lieutenant Rittichier the Air Medal in June 1967 for his role as the co-pilot of a helicopter that completed a dangerous rescue mission with the Mackinaw.

On November 29, 1966 Rittichier and his crew responded to a distress call from the West German motor vessel Nordmeer that had grounded on Thunder Bay Island Shoal in Lake Huron. The Mackinaw also responded to the distress call but prevailing weather conditions and the location of the stranded vessel prohibited her crew from effecting a rescue and they awaited the assistance of a Coast Guard helicopter.

Rittichier navigated the helicopter for 150 miles from Detroit with the "final 80 miles flown through

Lt. Jack Rittichier – a Coast Guard hero involved in a daring search and rescue operation to rescue sailors from the Nordmeer and hoisting them to the Mackinaw.

snow showers at 200 feet over the lake utilizing the shoreline for navigation," his commendation states. After locating the vessel, Rittichier established contact with her crew by radio. They indicated that they were stranded on the forward deck, exposed to the elements, had no power, and were in imminent danger. Rittichier then assisted the pilot in "maneuvering the helicopter and accomplishing the hoist of the eight crewmen from the Nordmeer to the decks of the Mackinaw" safely. The rescue was completed in a mere 22 minutes. Soon after the crew had been rescued, the Nordmeer broke apart and sank.

On June 9, 1968, Coast Guard Lt. Jack Rittichier was shot down over Laos, Vietnam while attempting to rescue a downed Marine fighter pilot. Rittichier was serving with a U. S. Air Force Combat Air Rescue detachment as part of a pilot exchange program. The crash site of Rittichier's helicopter, as well as his remains, was not discovered until the spring of 2003.

Rittichier was one of seven Coast Guard combat deaths during the Vietnam War. For his actions, Rittichier was posthumously awarded the Silver Star. The remains of Lt. Jack Rittichier, USCG, were recovered and returned to the United States. He was buried at Arlington National Cemetery on Coast Guard Hill Oct. 6, 2003.

Whenever the Mackinaw was on the scene, other vessels and agencies acceded to the giant icebreaker's presence as a command post and control center for whatever task was at hand. This occurred many times while standing by for icebreaking duties, whether in the Straits of Mackinac or elsewhere.

In February of 2001, the Mac was called to lead the search for a downed aircraft in northern Lake Michigan near Beaver Island. A chartered plane from Chicago carrying a young family went missing, and searchers had no idea whether the aircraft had ditched in the lake or crashed onshore.

The Mackinaw coordinated a search and rescue mission that involved multiple ships and aircraft. The effort took place during a major winter storm. The search effort resulted in the successful rescue of a mother and three children.

Over the years, the Mackinaw assisted sailors in the Chicago to Mackinac and Port Huron to Mackinac yacht races, aiding those who were dismasted, pitch-poled and knocked down and treating those with contusions, cuts, abrasions, sunburns, and heart attacks.

This 1947 Port Huron to Mac rescue mission resulted in towing the dis-masted yacht to Rogers City. The mast became unsecured in the rolling waves and was later cut loose by the Mackinaw crew.

In the early days, the fleet surgeon from the sponsoring yacht club generally rode on board the cutter.

"One year the fleet surgeon happened to be a pretty proficient sailor," recalled former Mackinaw skipper Jim Honke. "A Chicago to Mackinac sailor got hit in the head by the boom, and another sailor on board thought he was having a heart attack. A third guy wasn't experienced enough to sail the boat, he was basically a passenger.

'We put this young doctor into a small boat," Honke said, "and sent him over to the yacht. He revived the guy with the heart problem, treated the guy with the head injury and sailed the boat the rest of the way to Mackinac Island."

In the 2005 race the circumstances nearly repeated themselves.

A racing yacht, 12 miles west of the Mackinac

Claudia Simpson, the Mackinaw's first (and only) female corpsman, assists an injured sailor (lower left) on board his yacht in 2005 Chicago to Mackinac Race.

Bridge in bright sunshine and brisk winds, reported an injured crewmember and requested medical assistance from the Mackinaw, sailing not far away in Gray's Reef Passage.

"I left it in her hands," explained Cmdr. Joe McGuiness of the options he gave Health Services Tech 1st Class Claudia N. Simpson. "I told her we could have a helicopter pick him up, take her to the racing yacht, bring the injured sailor back to the Mackinaw, whatever she wanted to do."

"It sounded like he was in a lot of pain and was having some trouble breathing," said Simpson, the first and only female medical officer to serve on the Mackinaw. "Our crew sent me over in a rigid-hulled inflatable boat with a medical supply kit and I boarded to check out the situation. He was hurting, alright."

The sailor, a 47 year-old male from Charlevoix, suffered severe chest contusions after being injured on the vessel's foredeck. While working with the boat's spinnaker sail, the victim was thrown violently about while hanging on to the spinnaker pole and then thrown to the deck, striking some hardware associated with the rig.

"He said he was being thrown around like a greased pig," Simpson said. She recorded vital signs and started oxygen flow for the man, and determined that he was in too much pain for a transfer to the Mackinaw or to a helicopter. "The best place for him was right where he was. He was stabilized and comfortable. By the time we could have transferred him we had him on Mackinac Island."

Simpson stayed aboard the yacht and accompanied her patient on a downwind reach through the Straits of Mackinac under the Mackinac Bridge and then to Mackinac Island. She communicated vital signs and treatment plan by radio to the fleet surgeon while underway. The sailor was taken to Mackinac Island Medical Center for treatment.

Sometimes the emergency situtation took place on board the Mackinaw itself. Sailors from different eras have referred to a 1970s incident when a crewmember took his own life in a cabin onboard the ship as still affecting crews today. Even modern-day Coasties blame unusual events like a door opening for no reason or a missing piece of equipment as something "Willie" did. It is said that odd noises or other strange happenings are the result of his spirit returning to show his unhappiness with the ship.

Swimmer in the water…

Hoisting the victim up…

Coast Guard Helicopters frequently displayed their victim retrieval skills during Alumni Cruises, as they did in these 2004 photos.

Back to the chopper…

The Mackinaw, still dressed in white, outruns an ore carrier on yet another 1970s Great Lakes mission.

The Mackinaw waited offshore in the ice, pumping fuel oil to Mackinac Island residents while Bill Lorenz was with a crew ashore filling the tanks. The island's supply of heating fuel oil had run low in 1957.

On Mackinac Island, Lorenz made radio contact with the ship during fueling operations.

The J.P. Wells, in distress on Lake Superior with a broken rudder. The Mackinaw was planning on a 1956 Thanksgiving dinner onboard with dependents invited, but most families wound up with hamburgers or soup at home. The Mac headed up to assist the Wells, which was sailing in a circle. Eventually the cutter got alongside the damaged vessel and managed to tow it to a safe harbor.

Mackinaw crewmen ate their Thanksgiving Day dinner without their families in 1956, enroute to assist the freighter J.P. Wells. Mess cook W.R. Gable looks on during a break from the galley.

The Way It Was

Former crewmembers from the Mackinaw will readily spot things on board the vessel that are the same today as they were when they served in earlier eras. They will also point out many things that changed or became different due to remodeling, upgraded equipment or entirely new technology.

This is evident in photographs of the ship, both exterior and interior, taken at different times in the Mac's life. Different systems and the way they were used became standard operating procedure as changes took place on the cutter.

Even the preparations for sailing were noted as formal procedures done a certain way. Although the Mackinaw glides smoothly and effortlessly out of the harbor, there is a long list of readiness commands that take place long before she sounds her salute and departs.

Following is a list of preparations (circa 1970s) the officer of the day – or officer of the deck if at sea – would check off before the ship would sail:

- Four hours before sailing time the O.D. (referred to as the O.O.D. in modern times) makes sure that steps have been taken to energize the master gyrocompass. This is a delicate, intricate instrument that helps the ship find her way across the lakes. Other preparations follow but the next step does not occur until one hour

Nine Mackinaw crewmen crowded aboard for this 1957 lifeboat drill.

before the ship's departure.

- One hour before departure the O.D. makes sure that crewmen will be on the dock to throw off the ship's lines. He will now start the radar and set the radio watch. Now events proceed more rapidly.
- 45 minutes before sailing all hands must be mustered and absentees accounted for. The uniform for sailing is prescribed. Any of the Mackinaw's small boats in the water are brought aboard. The radar, which was started 15 minutes before, is now adjusted and tuned.
- At 30 minutes before departure the O.D. notifies all nearby ships of his intention to leave the harbor. All telephones links to shore are disconnected. The fresh water hose and electric cables to land connections are cut off. The ship's entire steering mechanism is checked. All navigation

Launching the Mackinaw's lifeboats required teamwork, balance, proper rigging and coordination.

equipment, whistles, alarms and signals are checked. The throttle, telegraphs and indicators are inspected and operated. Anchor windlasses and power winches on deck are actuated.

- 20 minutes before the ship sails the engineering, supply, deck, medical and operations departments report that the ship is in all respects ready to proceed. This includes major items like the boilers and engines being up to standards and minor items like ice cream being available for dessert. All visitors are now piped ashore.
- 15 minutes to go. The countdown continues as activities accelerate. "Set the special sea detail" is heard throughout the ship. All hands hustle to

In 1957 the Mackinaw's lifeboats were secured in a bunk arrangement with block, tackle and davits for launching.

their stations. Sound-powered phones are manned, used for intra-ship communication. Lines are singled up. The line-handlers, called for 45 minutes before, do their jobs. Messages are sent. The O.D. and his right-hand man, the quartermaster, shift the scenes of their activities to the bridge. The bridge, the heart of a ship at sea, now throbs with excitement.

- Only ten minutes until departure now. "The engines are ready," reports the engine control room. Next the bow detail checks in with "Anchors ready." More reports come in from various work stations around the vessel, such as "All singled up fore and aft" and "After steering manned and ready." 130 officers and crewmembers are acting as one, each checking their indi-

2004 aerial view of the Millard D. Olds Memorial Moorings, before the facility was upgraded to dock the original Mackinaw and the new Mackinaw, WLBB-30.

vidual stations and responsibilities. The controls are given a final test. The O.D. turns to the captain and reports, "Vessel is ready for getting underway, sir." The captain then replies, "Very well, take her out."

The sleek cutter moves slowly away from the dock. All is quiet and there is no noise or confusion. The ship moves so effortlessly that it seems simple and easy. She sounds her salute and departs.

Although recent crewmembers will note changes in procedure and computer-assisted technology, the basic formula for getting the Mackinaw underway remained the same into her final year of service.

Walter Holstad worked in the engine room from 1946 through 1949.

"It was all so nice and clean back then, everything was practically brand new," he declared at the 2004 Alumni Reunion. "It was a nice place to work. A little noisy, but not too bad."

The Mac received several overhauls and face-lifts throughout the years. Capt. Jim Honke oversaw a $1.7 million re-fit in 1982 at Sturgeon Bay, Wisc. that included the removal of the two giant seaplane cranes that were on the fantail. Included in the contract for renovation, a story in the Door County Advocate stated, were plans to update the auxiliary power plants aboard the ship. The four ship's service generators, of 1930s technology, were removed because of the difficulty in obtaining spare parts. They were replaced by three Caterpillar diesels and had new automated switchboards. In addition, four of the ship's

Crews quarters, 2003.

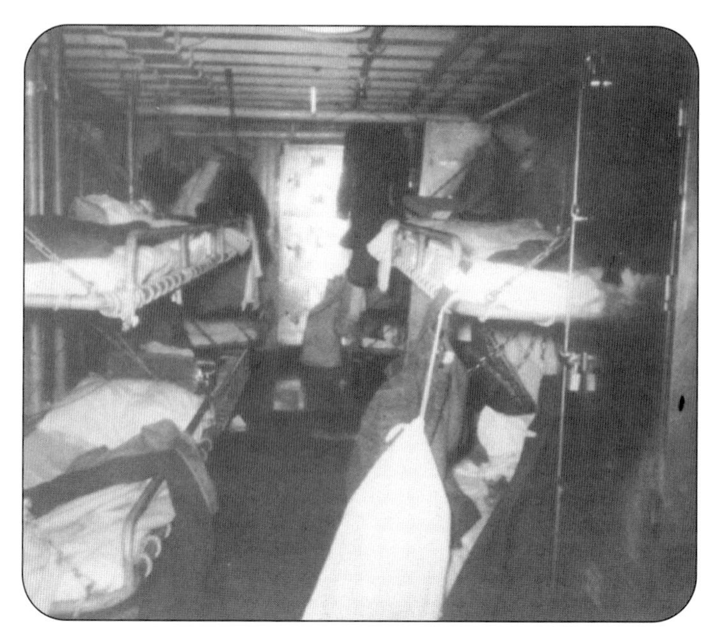

The crew's bunks were doubled up in 1967 when cadets came aboard for training.

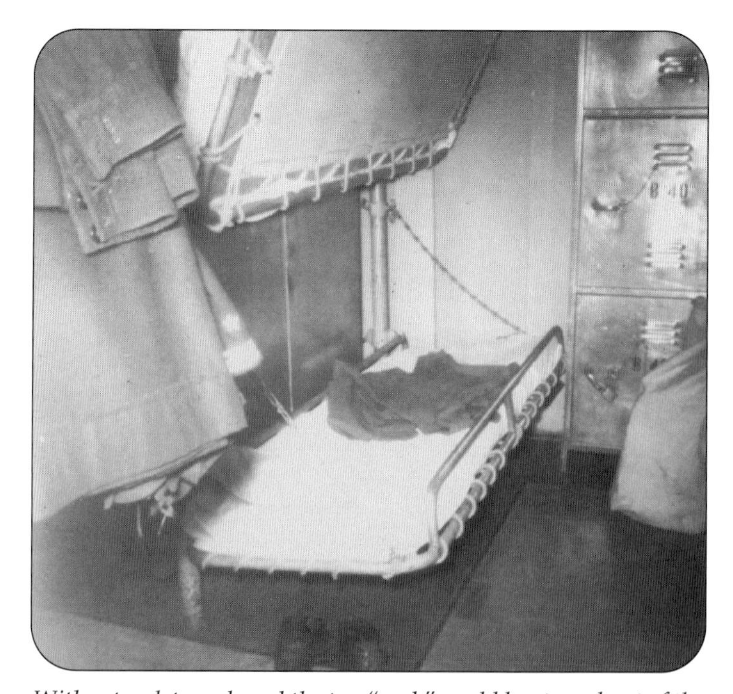

Without cadets on board the top "rack" could be stowed out of the way, giving a little more headroom.

"The highlight of my term was to climb the mast and bring down someone who panicked and could not come down on his own," said Robert G. Swanson, Jr. The Mackinaw was originally fitted with a crow's nest for a lookout to use while icebreaking or on a search and rescue mission. The perch featured a plastic glassed-in cubicle with a heated platform to warm the feet of the lookout. It was later removed.

"We were underway at the time," Swanson said. "The guy had climbed to the crow's nest and after he was relieved he got about six feet down the mast and froze. I guess I was the first one the chief saw and he told me to go up there and help him down. There isn't much more to it."

six propulsion generators were overhauled.

Another big project involved the addition of a lubricating oil purification system. Nearly one-third of the ship's fuel oil tanks were converted to fresh-water ballast, allowing the hull to ride lower in the water and increase efficiency in icebreaking work. The renovation of the fire control center was another area of extensive changes aboard the ship. The sewage incinerator was removed, and the starboard motor surfboat was replaced by a rigid-hulled inflatable type boat.

Due to cost, the Coast Guard scrapped the plans to renovate the berthing areas. Some of the original furnishings and equipment were still in use aboard the Mac in 1982, and this delay meant that because of the lack of proper facilities the Coast Guard could not yet assign any women to the ship. The crew remained all male until a subsequent remodeling in the late 1990s.

Hull-pitting from icebreaking work was also repaired, and the bearing staves around the shafts of all three propellers and the rudder bearings were all replaced. The shipyard even conducted a long-overdue inclining experiment, placing two-ton weights on the ship in order to find the vessel's new center of gravity. This was the first time the procedure had been performed since the ship was built in 1944.

When the Mackinaw was built in 1944, a gun turret was included in the design of the ship. According to David Lindahl, who was on the cutter from 1969 to 1972, it was left at Detroit.

"The Mackinaw originally had a gun turret but because the ship went into service after the war it never was placed on the ship. The last I heard it was left at the District Headquarters at Detroit," Lindahl said in 2004.

David Kaplan, on board during an earlier era, said this was due to an agreement between the U.S. Coast

Guard and the Canadian Coast Guard.

"We had machine guns in a machine gun rack back then, but we weren't allowed to have any guns on deck in the Great Lakes," Kaplan said. "It was a peace pact we made with Canada."

The Mackinaw was originally equipped with 40 M-1 rifles, 17 .45 caliber pistols, two Thompson submachine guns and two .30 caliber rifles. Since 2003, the vessel has hosted an M-60 emplacement.

The Mackinaw was a very expensive ship to maintain and upgrade, but re-fits and dry-dock periods occurred several times during her career. The engines were replaced. The bridge received new electrically heated windows during one remodeling, weather doors were replaced during another, and once the yard cut new windows in the quarterdeck shacks to improve visibility.

Each time the cost of a re-fit was considered, the Coast Guard was reminded of the price tag to actually replace the Mackinaw, which cost $10 million to build. In 1982, the replacement cost was estimated to be $60 million. The 1982 re-fit came after the ship's mission was revised, cutting the Mac's crew from 90 to 74. It had earlier been reduced from the 130 plank owners who were with the ship when it was commissioned. The overhaul came as part of a "recommissioning" that designated the ship in "transit mode" on a 24-hour basis, since it could no longer get underway at all times with full operational capability. In her ice-breaking mode, the ship at this time began operating on 12-hour days, and shut down during the night due to lack of manpower.

In the early 1990s the ship faced a real crisis with engine trouble, shaft trouble, and replacement parts issues that made the crew think they'd never get out of the yard and back on the lakes. Extensive repairs again made the vessel operable and back in service.

The ship was no stranger to the ports of

After Steering Panel

After Steering Apparatus

Marinette, Wisc., and Menominee, Mich. The Mackinaw was at Marinette Marine in 1992 for major repairs during a 54 day stay. It was an especially troublesome period for the vessel as once again parts could not be obtained for some systems and the ship seemed to be, for the most part, broken. Money seemed tight to fund these repairs and the Mac was again on the chopping block for decommissioning. But political pressure once more prevailed and the yard did a geat job with the repairs.

The work included new boilers, the power cables to the main engines were attuned and other overhauls were made. The public was given an opportunity to tour the icebreaker on Armed Forces Day in May, 1992.

It was not until many years later that the ship began receiving computer-programmed equipment for navigation and other functions. Equipment on the bridge changed significantly from 1944 until the Mac's last winter of operation in 2005-2006, but in many ways the systems remained the same.

Throughout the ship's service life, a bit of an urban legend existed that the Mackinaw could never leave the Great Lakes due to its width. Over the years, modifications in the lock system of the Seaway made this possible. Capt. William H. Fels, chief of the Eleventh Coast Guard District's Marine Safety Division, clarified that "While it couldn't easily leave the lakes, the Saint Lawrence Seaway accommodates vessels up to 76.11 feet, and the Mackinaw's beam is 74 feet. The reason it was lake-locked was due to its engineering systems, which assumes operation in a fresh water environment, not salt water. This means that the oil coolers are cooled by fresh lake water, whereas ocean-going vessels use a complex array of saltwater to fresh water to oil coolers in order to isolate the potential corruption of the engine oil with salt water. Even the make-up water for the Mac's boiler came straight from the lake! So, the Mighty Mac was destined to remain in the lakes, at least while being operated."

The Mackinaw proved her wide beam to be a non-issue on three occasions, traveling through the Welland Canal in 1967 and continuing as far east as Montreal for the Expo '67 World's Fair. The ship visited Toronto in July 1986 and again in 2002 with Cmdr. Jonathan Nickerson.

Close in the lock - The Mackinaw's wide beam often provided a tight squeeze in locks of the Great Lakes.

A long-awaited change in the duty rosters aboard the Mackinaw occurred in the late 1990s, when female crewmembers were finally able to begin serving on the ship thanks to re-modeled bunk and lavatory facilities. More than 30 female crewmembers served on the ship in its final years of service, and a like number of female cadets trained during the last summers the ship was commissioned. Several female ensigns and 1st and 2nd class petty officers also saw duty on the Mackinaw.

In the vessel's final years, the Mac had a female corpsman aboard who dealt with the crew's medical needs. Whether it was a routine health matter, a serious injury or just needing someone to talk to, officers and crew aboard the cutter knew they could count on "Doc," Health Services Tech. 1st Class Claudia N. Simpson, to pull them through.

"Doc" Simpson cared for all the above scenarios and plenty more during her two years aboard. Newcomers aboard the giant icebreaker quickly learned that a visit to Simpson's office was a straightforward, no-nonsense experience that just happened to be tendered with a lot of caring and concern.

"They've come to me with everything from a blister to needing stitches to being homesick," Simpson said while updating inoculation records for the Mackinaw's crew in 2005. "I think a lot of the younger ones kind of looked at me as a mom. I really liked the duty."

During her time aboard the Mackinaw, Simpson also dealt with varied medical emergencies that involved a transfer to a hospital including "some heart problems, a near-amputation, and a guy who impaled himself with a five-inch screw," she said. "I'm the first female corpsman on the Mac and I'll be the last female corpsman on the Mac. I'm so proud to be on this ship; we're the last crew and it's really an honor."

Simpson's career in the Coast Guard began through the encouragement of her father and her search for

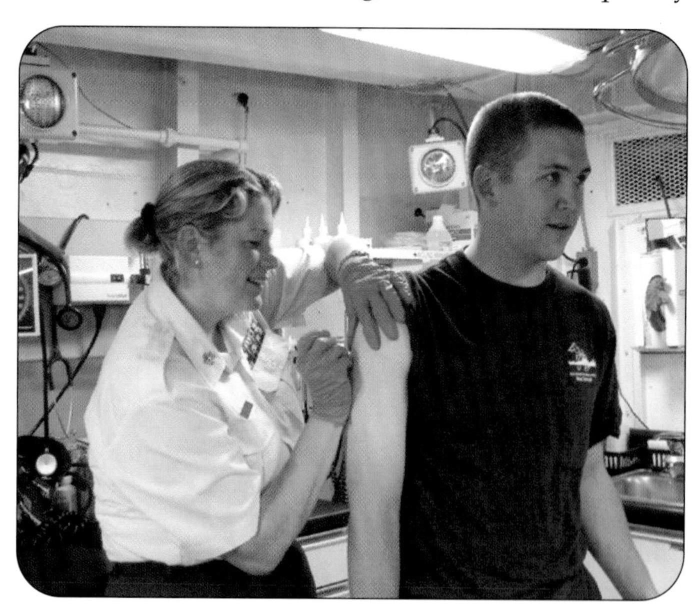

"Doc" Claudia Simpson inoculates Ensign James Conner after checking his medical chart.

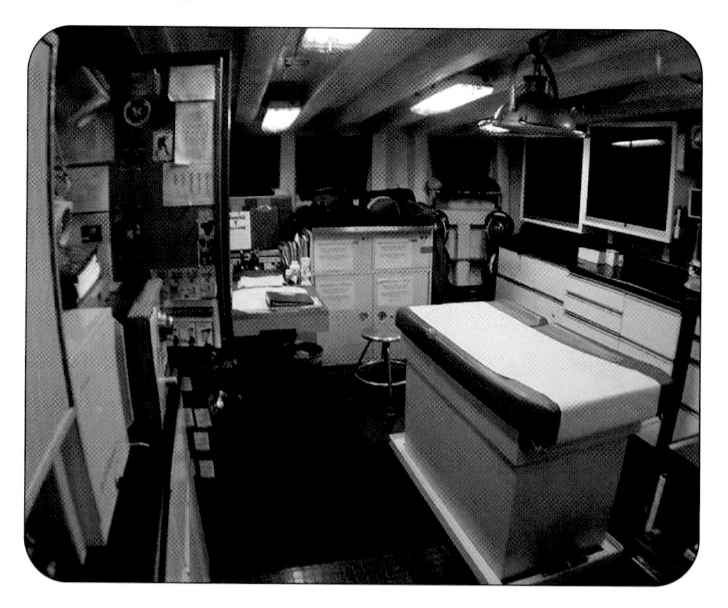

Sick Bay, where medical corpsmen treat and maintain the health of the crew.

work in public service.

"I wanted to do something to help people," she recalled. "It started as a two-year trial and it became a career. My mom is in health care and it's been a great fit for me. I've had great commanding officers and everybody's treated me well. It's like an extended family aboard this ship.

"I've gotten to know some fantastic people," she exclaimed. "The Coast Guard focuses on diversity – women and minorities – so everyone feels important, everyone feels equal. I tell incoming female recruits to respect other people and avoid putting yourself in a bad situation."

Simpson anticipated receiving her five-year cutter pin in December, 2005, and was justifiably proud of her honor.

"Not many women in the Coast Guard who are corpsmen have one," she smiled.

The Mackinaw's brow at the Cheboygan dock in 2005.

The deck crew throw shotlines from the bow to tie up the Mackinaw as it arrives in Cheboygan in 2005.

Mac crewmen tie up the stern lines.

Door to the ship's supply store, 1978

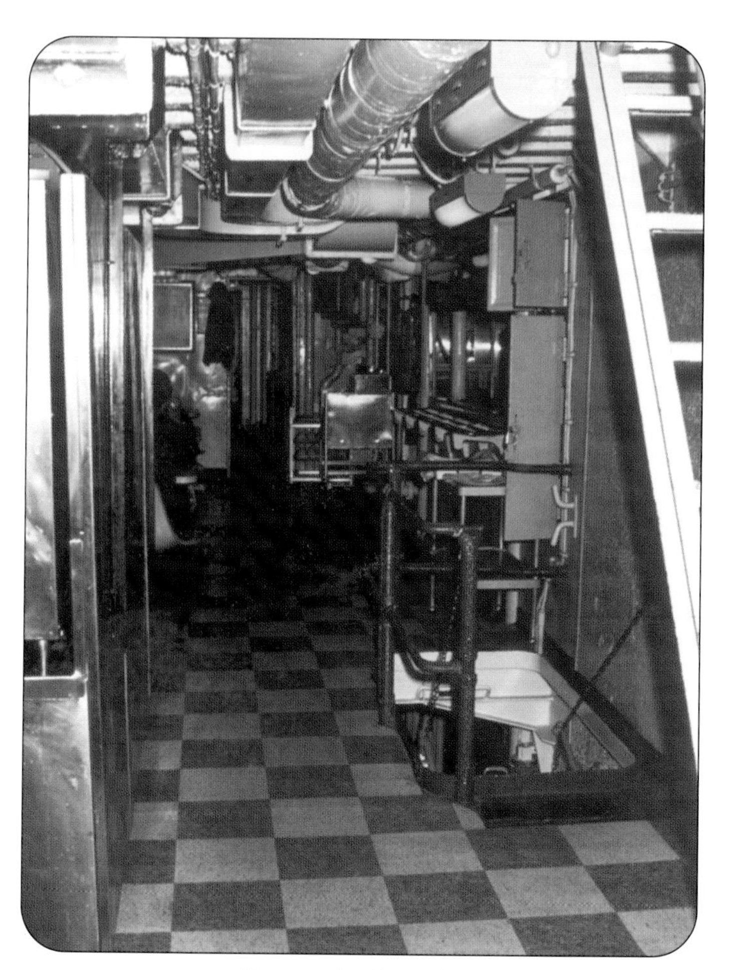

Passage to the galley

The ship's office and library later became a station with computer terminals for crew use.

The ship's workshop contains tools to handle many repair jobs.

It took more than elbow grease to replace Boiler No. 1 in the engine room in 2005. Petty Officer 2nd Class Brian Evans (left) stands by as pipefitters Ed Torres (kneeling) and Charles Bare from the U.S. Coast Guard's Curtis Bay Shipyard, near Baltimore, Md., tighten a fitting with an enormous wrench. The new 1,600-pound boiler was inched through the decks and passageways of the Mackinaw by two rigging specialists, who also traveled from Maryland for the job of replacing the 15-year-old boiler. The Schwartz Boiler Shop assisted in the project with use of its truck and crane for onloading and offloading.

The commanding officer's cabin as it looked in 1979.

The mess deck has been the scene of thousands and thousands of meals since 1944. This 1979 photo shows the older dining furniture that was replaced in 1982.

The enlisted crew's head as it appeared in 1979, prior to a major re-fit in 1982 that remodeled these facilities but did not accommodate female crewmembers, not rostered on the ship until the 1990s.

A 1980 stern shot of the Mackinaw breaking ice, showing the ship's fantail deck, helicopter pad and seaplane cranes.

Side view of "Big Bertha," the Mac's giant towing winch with three-inch cable.

Engine room No. 2, known as "two-space.".

One of the Mac's three boilers.

Forward view of the towing winch.

No. 2 Alpha generator in "three-space," used for emergency power.

Action on the bridge with pilothouse controls near all ahead full on both aft shafts and the bow shaft stopped. Cmdr. Jonathan Nickerson is seated in the captain's chair.

The watch stander used a bridge wing enclosure for ice breaking, docking and close maneuvers.

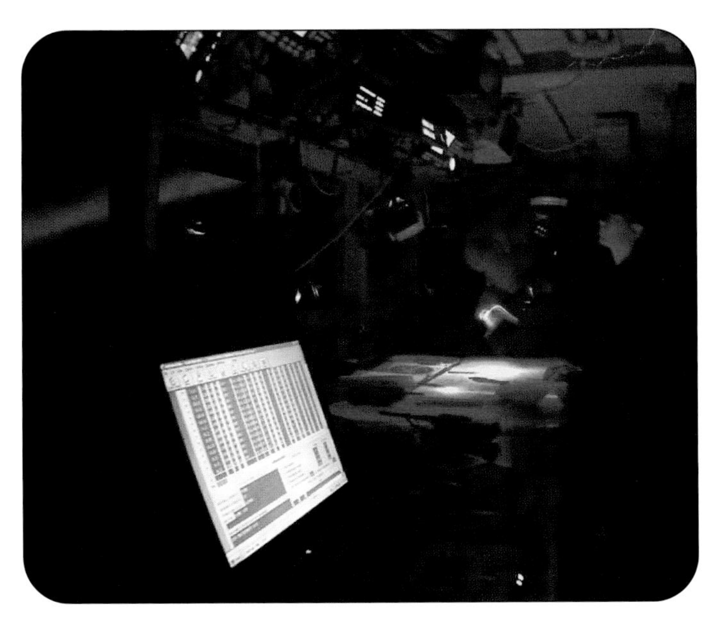

Nightwatch on the Mac's Bridge under the red glow of night-vision lighting.

The ship's wheel and steering station.

Dimensions of the USCGC MACKINAW (WAGB-83)

MACKINAW WAGB 83......Icebreaker, Heavy
(Total of one in service of this class)
Builder ....Toledo Shipbuilding Co. Toledo, Ohio
Keel laid ....March 20, 1943
Launched ....March 4, 1944
Commissioned........December 20, 1944

GENERAL CHARACTERISTICS:

Length Overall ....290 feet
Length between perpendiculars........................280 feet
Breadth, Extreme.........74 feet, 4 inches
Full Load Displacement ....5,252.4 tons
Mean Draft, at Full Load Displacement.....19 feet, 2 inches
Propulsion......................Diesel-Electric
Personnel: Officers ....13 Officers
Crew ....64 Enlisted
Anchor.......................Two 6,000 pound Bower Stockless
Chain....Two 90-fathom lengths 2-inch diameter, wrought iron stud link
Breaking strength: 225,000 pounds

STORAGE CAPACITIES

Diesel fuel.....................346,910 gallons
Potable water ....40,200 gallons
Heel/trim ballast water. .345,828 gallons

Heights Above Waterline (at Full Load Displacement)

Foc'sle . . . . .20 feet, 10 inches
Fantail . . . . .9 feet, 4 inches
Bridge . . . . .29 feet, 10 inches
Flying Bridge . . . . . . .37 feet, 2 inches
Mast Head . . . . .105 feet

SMALL BOATS

One 25-foot Motor Surf Boat (MSB)
One 7-meter Rigid-Hulled Inflatable Boat (RHIB)

ENGINEERING CHARACTERISTICS:

MAIN ENGINES

Number of Engines 6
Type of Engines . . . . . .Diesel, 2-cycle opposed piston Fairbanks-Morse
Horsepower
Continuous . . . . . .1750 HP (each)
4-Hour rating . . . 2000 HP (each)
Number of Cylinders 10
Revolutions per minute
Continuous . . . . .750 RPM
4-Hour rating . . . . .810 RPM
Starting Method . .Air Starting System

PROPULSION GENERATORS/MOTORS

Number of Generators........................6
Manufacturer.................Westinghouse
Volts ....900 V
Revolutions per minute ....810 RPM
Number of Motors: Forward 1
Aft. 2
Manufacturer.................Westinghouse
Horsepower (each)
Forward .... 3300 HP
Aft......................................5000 HP
Volts ....900 V
Revolutions per minute:
Forward....................175-200 RPM
Aft.............................136-170 RPM

PROPELLERS

Number: Forward 1
Aft 2
Type Solid Steel
Weight: Forward . . . . 7.7 Tons
Aft . . . . .10.7 Tons
Number of Blades 3
Diameter Forward . . . . .12 feet
Aft . . . . .14 feet
Shaft Diameter Forward .13 1/4 inches
Aft . . . . .16 1/2 inches

SHIP'S SERVICE GENERATORS

Number of Engines.............................3
Type of Engine....................Diesel, V-8
Manufacturer ....Caterpillar D379
Horsepower...............................650 HP
Revolutions per minute................1300
Method of starting............(2) Air Start, (1) Battery start
Number of Generators 3
Manufacturer ....Kato Engineering
Voltage ....450 V
Current....................................674 AMP
Frequency ....60 Hz
Power ....420 KW
Revolutions per minute.......1200 RPM
Type.........Three-Phase AC Generators

The Most Popular Ship In Town

The Mackinaw was always the star attraction for any festival or event it attended and drew attention in every harbor it visited. When the ship docked, people showed up and wanted a tour.

They were usually accommodated, for the Mac's crew proudly showed off their ship whenever they had the chance, whether it was in Grand Haven, Cleveland, Chicago, Sault Ste. Marie or anywhere else the lines were tied. They were often aided by the local Coast Guard Auxiliary squadron.

This spirit extended to many crewmembers while homeported in Cheboygan, who kept up the same goodwill there they were famous for all around the Great Lakes.

The Mackinaw held a popular Halloween "Haunted Ship" party for many years, decorating the vessel for tours by children and their parents. One set of passageways would feature mildly-frightening sights for younger children, while another area of the ship was set aside for older guests who really wanted to be scared. Sometimes people arriving in the parking lot at the Millard D. Olds Memorial Moorings would hear screams coming from the ship, followed by raucous laughter by crewmembers who especially delighted in scaring young ladies. The crew pulled it off each year and provided a safe Halloween party that allowed everyone who visited the ship to leave smiling.

Blood drives, food drives and other benefits for Cheboygan often took place aboard the ship and at various locations around town. Crewmembers also held holiday parties for children of Mackinaw families, making them feel special with Christmas parties, Easter egg hunts and summer picnics.

The Mackinaw's benevolence extended to other Great Lakes ports as well. The ship annually attended

Mackinaw crewmen preparing for the 2004 Halloween "Haunted Ship" party.

the Coast Guard Festival in Grand Haven, and made appearances at Traverse City's National Cherry Festival and many other special events in the summer months.

The Mac provided a safety escort each July for the Chicago to Mackinac yacht race, shepherding the participants across the 333-mile expanse of Lake Michigan and under the Mackinac Bridge to the race finish line at Mackinac Island. There, a cannon would fire announcing each sailboat's arrival.

For many years the ship also handled the same duties up Lake Huron for the Port Huron to Mackinac race, until smaller cutters eased the load. Unless the Mackinaw was in the shipyard undergoing repairs, the Chicago race was a regular event always on the ship's calendar.

First, the Mac served as a host to several hundred spectators who watched the start of the race one mile east of Chicago before heading out to act as the safety watchdog over the fleet of 300 yachts.

The Mackinaw served as a foggy backdrop for this 1986 shot of the National Cherry Festival Queens at Traverse City.

Starting at 3 p.m. on Sunday of the race weekend, the Mackinaw would begin to receive radio calls from each class of boats in the fleet, stating their location on a grid in the race scratch sheet. By this time the leaders could often be too far ahead and out of radio range. In later years, this job was performed by computer software and handled by the race committee, who logged the location of each yacht through the use of cell phones and GPS transponders.

Once the Mackinaw reached the Manitou Islands off the Michigan coast near Traverse City, the ship would sometimes anchor and wait for the fleet to pass. There the ship again would monitor the boats' locations to provide additional information to the Chicago Yacht Club Race Committee stationed at Mackinac Island. When the fleet had passed the

The Mackinaw, surrounding by sailing yachts in preparation for the 2005 Chicago to Mackinac Race, the Mac's last.

Manitous, the cutter moved on to Gray's Reef Passage where it anchored and tried to account for every boat.

Some years the race would go by briskly without incident, while during others the yachts ghosted along, becalmed by quiet seas. There were years where violent storms rocked the lake, usually in the middle of the night, and boats were dismasted. The Mackinaw handled man-overboard situations and treated sailors with various injuries. The Mac would act as the safety net for yacht crews that would be awake for two days straight and find themselves in unsafe conditions. Most of the time it was pleasant duty with relatively few distress calls.

The ship has enjoyed a special relationship with the city of Chicago, hosting an annual summer "Purple Heart Cruise" for war veterans and in later years returning in December as the "Christmas Tree Ship."

"Kup's Purple Heart Cruise," the brainchild of Chicago newspaper columnist Irv Kupcinet, could well be the most fun six-hour trip the ship took each summer for 50 years. America's war veterans from all eras and those in active duty were invited aboard to

Chicago to Mackinac races were one of the ship's most popular summer duties as the Coast Guard's ambassador on the Great Lakes.

receive a well-deserved pat on the back for protecting the freedom of its citizens.

From the minute the buses unloaded until the Mackinaw returned to the pier six hours later, there was constant entertainment going on. Musicians provided everything from the big-band sound to light jazz to rock 'n roll, as Mackinaw was transformed into a floating carnival. Over the years the ship carried magicians, clowns, a palm reader, models, strolling troubadours and a variety show reminiscent of a USO tour. There were meals and beverages served constantly at various locations on the ship, and veterans from four Chicago-area Veterans Administration hospitals and other military installations were given the hero's treatment they deserved.

Kupcinet began the cruise in 1945 as World War II was drawing to a close.

"During the war, especially in Chicago, the servicemen had been treated like kings," Kup noted in a 1985 story in VFW Magazine. "I was afraid that as the war ended they would be forgotten."

Through his newspaper column, Kupcinet convinced the people of Chicago to donate their money, merchandise and talent to make the day a success for the 500 or more veterans who were brought aboard each year.

The Chicago Yacht Club, grateful for the years the Mackinaw escorted its fleet of racing sailboats up Lake Michigan, heard stories of veterans who couldn't tolerate the bright sunshine on the Mac's fantail stern deck. After suggesting a white canvas cover, the club picked up the tab for the giant weather shield that protected cruisers from sun, rain and winds for many years.

The cruise continued for 50 years until 1994, when Kupcinet's health began to decline. He died in 2003, and the city of Chicago flew its collective heart at half-staff. One man did so much for so many, and used the Mac as the centerpiece.

Soon after the Purple Heart Cruises ended, the Christmas Tree Ship visits began under the direction of Cmdr. J.H. Nickerson. Each December since 2000, Mackinaw crewmembers headed out into the Northern Michigan forests to cut Christmas trees for delivery to Chicago for needy families as part of a re-enactment of the city's famous Christmas Tree Ship.

The original Christmas Tree Ship was the Rouse Simmons, a three-masted schooner that brought Michigan evergreens to Chicago every year from 1877 until it sank in an early winter squall on Nov. 22, 1912. Chicagoans became accustomed to purchasing their

Mackinaw crewmen cut Christmas trees at Northern Michigan farms, bundled them, and transported the load to the ship for the trip to Chicago as the "Christmas Tree Ship".

wreaths and trees this way as a festive start to the holiday season. Eventually, a number of trees were brought along specifically for needy families who couldn't afford a tree. Tragedy temporarily ended the tradition when a 1912 Lake Michigan blizzard claimed the Rouse Simmons, its skipper Herman Schuenemann and all 16 hands, lost with more than 5,000 trees.

"Everybody goodbye," read a note in a bottle found by the U.S. Lifesaving Service – predecessor to the Coast Guard – as it searched for the wreckage of the Rouse Simmons. "I guess we are through ... God help us."

Mackinaw crewmen load hundreds of Christmas trees aboard the ship, continuing a tradition begun by Cmdr. Jonathan Nickerson in 2000 as the "Christmas Tree Ship" carrying Christmas trees to needy families in Chicago. A large quantity of Christmas trees were also provided for Northern Michigan families from the Cheboygan dock.

Historians believe the note was written by Schuenemann, known to Chicagoans as Captain Santa. Others acknowledge the discovery of the note, but claim it was a merely a cruel hoax played up by newspapers of the era.

"When the Rouse Simmons sank, it was a huge loss because the ship had become a tradition," Jane Munch, a Coast Guard spokeswoman, told The Alpena News in 2000. "The whole city welcomed Schuenemann every time he pulled into the harbor."

As the storm surged, ice started building heavily on the ship's deck and the freshly-cut trees that were lashed across it. The Rouse Simmons was seen briefly the next day off the Door Peninsula in Wisconsin. The crew was flying distress flags, and the ship's sails were tattered, but rescuers couldn't reach them.

The 200-ton schooner sank in 180 feet of water off Two Rivers, Wis. Afterward, evergreen trees occasionally floated to the surface in the area of the wreck, and a diver found the wreckage in 1971 and claimed there were remnants of bundled trees still aboard.

In the fall of 2000, a group of maritime-history buffs approached the Mackinaw about retracing the route of the Rouse Simmons. Nickerson said he and the 75 members of the Cheboygan-based cutter readily agreed.

"This is an opportunity for us to do something for the community of Chicago and at the same time provide positive exposure for the Coast Guard," Nickerson told the Alpena newspaper.

When the Mackinaw left Cheboygan each December for the 24-hour trip to Chicago, the cutter often encountered some snow showers en route, usually traveling within 20 miles of the Rouse Simmons' watery grave.

Upon reaching Chicago, the Mackinaw encountered a welcome as warm as what Schuenemann and his crew must have received those many years ago. Volunteer organizers would join crew members in stringing colored Christmas lights from the cutter's bow to its stern.

The voyage also gave the Mackinaw's crew an opportunity to take part in offshore training and onshore community service. Crewmembers conducted tours of the Mackinaw through the weekend.

"The cutter has always been a great draw because she is a pretty incredible sight," said Kevin Pilarski of Chicago, one of the organizers of the Christmas Tree Ship's voyage. "As folklore has it, passage on the Great Lakes can be rough in the months of November and December. The cutter denotes and represents safety on the Great Lakes, and we should all appreciate her."

Until the Mackinaw's retirement, the concept of Chicago's Christmas Ship remained active as a charitable event organized by the unified marine community of Chicago in cooperation with the Coast Guard.

Part of the Mackinaw's fall routine involved work-up trips in preparation for the icebreaking season. In conjunction with the work-up trip to Chicago, the Mackinaw was able to transport trees to be donated to Chicago's disadvantaged families.

In addition, the crew of the Mackinaw distributed more than 100 trees annually to disadvantaged families throughout the Northern Michigan area. The Cheboygan post of the Michigan State Police, the Cheboygan County Sheriff's Department, the City of Cheboygan Department of Public Safety, the Salvation Army, and other Northern Michigan Coast Guard units all worked together to distribute the trees. Cheboygan-area Christmas tree farms would often donate the trees to the cause, and the Mackinaw's crew would handle the work detail with chainsaws and equipment for banding the trees together for the journey. The Mac brought trees to Emmet County in 2005, stopping at Harbor Springs.

Due to the ever-increasing frequency of repairs over the years, the Mackinaw also developed a strong Wisconsin connection and did not forget those cities when ice-breaking time came around.

Larry Ebsch reported in the Marinette/ Menominee Herald that the Mackinaw cleared many paths in clogged Green Bay ice to enable the Ann Arbor car ferries to reach Menominee during the 1940s and 1950s when the ferries moved passengers and many of the products that were manufactured there.

One of those critical times was in February of 1953 when the ferries Wabash and No. 5 were snagged in three feet of ice for 36 hours near Sherwood Point. Any delays in Great Lakes shipping during the harsh winter months created problems for local industries that depended on the car ferry line as a major shipping source.

The Mackinaw was moored at the Menominee Marina for the colorful "Blessing of the Fleet" ceremonies as part of the 1949 festivities. The ship returned to Menominee in July 1966 for the same event, which by then was re-named the "Blessing of the Watercraft." An "International Avenue of Foods" was another highlight.

One of the early memorable assignments for the Mackinaw came in the summer of 1945. Germany had already surrendered in May, and Japan was only one month away from collapsing. For the first time in the history of the Coast Guard, cadets from the Coast Guard Academy in New London, Conn., trained on the Great Lakes.

Two groups of about 110 cadets were assigned to the Mackinaw for 18 days of training. According to the Coast Guard, it was the first time since the schooner Dobbin made the first cadet cruise in the Atlantic Ocean in 1877 that cadets trained anywhere but in salt water. The training mission sent the cadets

1984 docking at Grand Haven.

to the bulging industrial plants in the Detroit area, the Sault Ste. Marie Locks in Upper Michigan and the ore docks at Duluth, Minn.

Cadet training became a regular summer activity in later years.

During the summer of 2001, the Mackinaw was honored by the Detroit 300th Anniversary Celebration when the vessel served as the lead ship for the historic Tall Ship Parade. It was surely not the first time the Mac has sailed in front of the pack. Whenever the cutter departs Cheboygan or any other port, a line of boats forms pretty quickly to follow her out of the harbor. The Mackinaw has always been the leader and centerpiece of any nautical parade or celebration she has attended.

In 2004, Food Service 2nd Class Joseph Kraft, a cook on the Mackinaw, had an idea that some food distributors he worked with might be in a position to donate food items to a project known as "Christmas Wishlist" sponsored by the Cheboygan Jaycees. Kraft was a prospective member of the civic organization at the time.

The hunch paid off, and Kraft coordinated delivery of boxes of non-perishable items to Jaycees representatives courtesy of food suppliers who regularly stock the Mackinaw with ingredients for meals.

"In all there's more than $2,000 in goods," Kraft determined. "We have chickens, turkeys, hams, anything they could make for a traditional Christmas dinner. There are even kids foods and breakfast foods."

Kraft, originally from Rochester, N.Y., purchased the ship's stores locally and found his business partners more than willing to help out.

"It's a chance to use the Mackinaw's buying power to help the community in a way we hadn't looked at before," he said. "The ship is a good customer for these companies and they were willing to get involved to help us help people in our home port."

According to Kraft, military food for the Coast Guard used to be purchased in bulk through a Philadelphia warehouse and everybody ate the same things most of the time.

"Military food service has changed over the years," he explained. "I can give our crew a taste of local favorites or develop menus based on their likes and dislikes. If they like it, we'll have it again."

In 2004, Kraft and other cooks onboard worked under the command of Food Service Chief Terry Sorenson. While the Mackinaw is underway the mess crew prepared 80 meals four times a day to cover six watches. Kraft said it's a never-ending process.

"We tend to serve more hot foods and heavier-type meals during the winter icebreaking season," he noted. "Most of our meals are made from scratch, like soups and things like that due to the quantities involved. We want to keep them happy."

Some meal themes took on a regular schedule during this time, such as the lunchtime "Taco Tuesdays" and the ship's famous "Cheboygers" that were always served on Wednesdays for lunch.

The Mac departs the harbor at Buffalo, N.Y. in 1986.

The fat, football-like shape of the Mackinaw is evident as the ship breaks ice near Port Huron, Mich.

The Mackinaw's parade unit at Grand Haven in 1975 for the Coast Guard Festival.

The Mackinaw posed in full dress colors for the Coast Guard Day celebration at Grand Haven in 1957.

Ceremonies to honor the memory of the U.S. Coast Guard cutter Escanaba at the same Grand Haven dock in 2005, 48 years after the photo on the previous page was taken.

The ship's final visit to Grand Haven's Coast Guard Festival may have topped them all with a fireworks grand finale in 2005. Note the number of small craft sharing in the celebration with the Big Mac as the centerpiece.

They lined up to see the Mac in Toledo – like they did everywhere else – for tours of the ship. Young or old, military or civilian, male or female the icebreaker was a popular attraction for many Great Lakes festivals.

The Mackinaw is shown here in 1988 downbound from Lake Superior, passing under the Soo railroad bridge before entering the Soo locks.

A night view, under searchlight, of the same Soo Locks entranceway as at left.

This 2003 docking at Group Sault had the brow placed in plenty of snow, as usual.

A 1970s - era photo of the Mackinaw, upbound in the St. Mary's River for the Soo Locks and Lake Superior icebreaking duties.

Moored along the lock wall in Sault Ste. Marie, Mich.

Entering the Poe Lock at Sault Ste. Marie, sharing the space with a tug boat.

The Mackinaw is shown here docked near Detroit's Renaissance Center in 2003.

Approaching a lake freighter for breakout in the 1980s near Marysville, Mich.

Departing the Maumee River near Toledo, the Mac's birthplace.

The Nine Lives of the Mackinaw

It is indisputable that Cheboygan, Mich., has always been the homeport of the Mackinaw from that first trip up the Cheboygan River in 1944. There is also no question that other cities wanted the ship and on many occasions tried to get her. Sometimes it appeared that even the U.S. Government wanted the Mackinaw out of commission.

It is true that Cheboygan more or less snuck the Mackinaw right from under the city of Milwaukee's nose while plans were being made for the ship's construction. Ever since, it seems, other communities thought they could steal the Mac for their harbor or downtown waterfront.

A plaque in the crew's mess on board the ship declares that the Mackinaw was "named for Mackinaw City, Michigan." The village just 15 miles north of Cheboygan was always a friendly port for the icebreaker, but a change in command once brought a new captain who figured the ship would be better off staying in the vacation community.

"In 1956 it seemed that in all those years Cheboygan was apparently the 'temporary' homeport of the Mackinaw," former crewmember Jack Eckert recalled. "The new captain wanted to move to Mackinaw City where the docking facilities were better. I guess he didn't like the sight of our big white ship docked behind a coal pile. The Mackinac Bridge was near completion and the ferry boat docks there were much better and soon would become available for use. Needless to say the townspeople were not only mad at the skipper but at us too."

The Mackinaw City move never advanced any further, but from time to time rumblings were heard that the Mac might soon be on the move elsewhere.

In 1962 the city of Port Huron made what an editorial in the Grand Haven Tribune referred to as "par-

A 1962 petition drive to save the Mackinaw utilized schooboys who went door-to-door soliciting signatures and donations for petitions sent by telegram to Washington, D.C. legislators.

ticularly aggressive" moves to try and squeeze the Mackinaw away from Cheboygan in favor of a new berth near the Bluewater Bridge.

The Coast Guard estimated that $400,000 needed to be invested in property improvements if the icebreaker were to remain at Cheboygan. The Cheboygan Daily Tribune quoted U.S. Rep. James G. O'Hara, D-Michigan, as saying that the money could be saved simply by moving the ship to Port Huron where docking and housing facilities were already in place.

"There are plenty of facilities, nothing need be built," O'Hara said of Port Huron, located within his congressional district. "Most of the crewmen won't bring their families to Cheboygan. There's plenty of housing available at Port Huron."

U.S. Rep. Victor A. Knox, R-Michigan, disagreed but said the Coast Guard could have had the money for the moorings project long ago if they'd asked.

"I believe there is adequate housing at Cheboygan," Knox said. "I've heard of no shortage. I have constantly asked the Coast Guard to make a request of the budget bureau for the money it needs. Permanent moorings could have been built for $280,000 in 1959 or 1960, but now they're talking about $400,000."

Grand Haven sentiment favored keeping the ship more accessible to breaking ice in Lake Michigan. Port Huron boasted close proximity to Detroit waterways and the lower lakes, with the upper lakes just hours away. It all sounded glamorous and convenient for Port Huron, but the idea was quite depressing for the community of Cheboygan.

The Cheboygan Chamber of Commerce, with endorsement from the chambers of Sault Ste. Marie and Petoskey, began a campaign called "S.O.S." to "save our ship" for Cheboygan. Mrs. Betty Jane Minsky, secretary/manager, spearheaded a drive along with Chamber president Victor Leonall to canvass the city for signatures and 25 cent donations to send telegrams to U.S. Senators Philip Hart and the commandant of the Coast Guard. Businesses and service clubs got behind the effort and donated checks, and a corps of nine young men were commissioned to solicit the signatures at a day's wage of $4. The boys were under the direction of Cass Lempke and included John Gardner, Charles Couture, Larry Bucalous, John Illig, Ron Denmur, Dale Gauthier, Steven Williams, Ray Drake and Gregory McKinley.

The Cheboygan Daily Tribune endorsed the cause and urged people to take action. The group raised more than $243 and sent telegrams signed by 677 people.

The wires stated, "We, the residents of Cheboygan, strongly urge that the icebreaker Mackinaw be retained here in that this is the center and most ideal location. We pledge ourselves and our facilities to provide a pleasant association between Cheboygan and the personnel of the ship."

In the days that followed, a delegation of 13 citizens met at Sault Ste. Marie with Admiral Edwin Roland, the Mackinaw's first commanding officer, to discuss the issue. Attending were Steve Majestic,

The Mackinaw arriving at Duluth in 2001.

Archie Barnich, Joe Van Antwerp, Lloyd Guenther, Arnold Jorgenson, Joe Louisignau, Victor Leonall, Betty Jane Minsky, John Nieman, Mr. and Mrs. Al Gates, and Mr. and Mrs. Nick Hile.

Roland told the group that besides the matter of the $400,000 needed to build the Cheboygan dock complex, "local conditions (in Cheboygan) have served to create a morale problem among the personnel of the Mackinaw. This is not a desirable situation, and has served to create domestic and disciplinary problems which have had serious adverse effects on the operational efficiency of the vessel."

At issue was the lack of suitable housing in the area and few inducements for crewmen to bring their families. High rental prices for crewmen were also cited.

Cheboygan was supported by the Sault Ste. Marie Chamber of Commerce and the Petoskey Chamber of Commerce, and both groups wrote letters of support to congressmen, senators and Coast Guard officials.

"Cheboygan needs help instead of a push backward," the Petoskey News-Review quoted a Chamber official there as saying.

Roland ordered a study conducted to review the advantages and disadvantages of several Great Lakes ports as a possible home base for the Mackinaw. He said indications were that Port Huron enjoyed several advantages over the other ports.

Next, a two-man delegation visited Cleveland to solicit support for Cheboygan. Mayor James Muschell and Steve Majestic called on retired Admiral James Herschfield, president of the Lake Carriers Association, and Vice Adm. George H. Miller, the commanding officer of the 9th Coast Guard District. They took along a proclamation from the Cheboygan City Council to plead their case that the city was depressed economically and simply could not afford to lose the cutter's payroll. The influence worked and the change didn't happen.

Any dissatisfaction with the Coal Dock's dirty location – and there was plenty – seemed to end when the family of Millard W. Olds donated the land for a new mooring facility across the Cheboygan river. The Coast Guard at last had its own space to base the cutter.

However, the dock was located in the turning basin of the river and was sparsely furnished. Over time the Mac's "permanent" home began to need major improvements. Maintenance costs began to escalate until the Coast guard decided in 1973 to build new moorings at Cheboygan to provide for the vessel. The decision was made to locate the dock out on the river, adjacent to the turning basin. The improvements would be costly, and would include a community building for the Mackinaw's crew, parking and recreational facilities.

It was then that two more Michigan congressmen became involved in a battle over where the icebreaker should tie her lines.

U.S. Rep. Guy Vander Jagt, R-Luther, started the fight by offering an amendment to a Coast Guard appropriation bill, deleting $1.6 million allocated for building the new dock and amenities. His attempts to save money came immediately after he added $600,000 to the bill to re-open Coast Guard search and rescue stations at Manistee and elsewhere. He proposed a move to instead dock the ship in his district at Grand Haven, where there were already excellent moorings for a ship,

U.S. Rep. Guy Vander Jagt

completed in 1972.

His amendment, said Vander Jagt, would "enable the taxpayers of America to save $1 million."

U.S. Rep. Philip Ruppe

U.S. Rep. Philip Ruppe, R-Houghton, whose district included Cheboygan, fired back that Vander Jagt was attempting to commit "Great Lakes piracy."

The resulting furor caused the U.S. Coast Guard's 9th District office in Cleveland to assemble a report on the establishment of a "permanent" home for the Mackinaw.

Again.

Citizens in Cheboygan were disenchanted to learn that the Coast Guard had considered moving the Mackinaw only 13 months after the ship first came to town. According to the report, published in the Grand Haven Tribune, "numerous attempts have been made to change the home port of the Mackinaw since the ship was based there after commissioning in 1944."

It was also pointed out by reporter Clarence Poel that the location of the ship near the Straits of Mackinac – originally considered of prime importance – had drastically decreased in necessity with the construction of the Mackinac Bridge linking the Upper and Lower Peninsulas of Michigan.

"The work of the cutter Mackinaw is now more diversified than ever which would not make a port location near the Straits of Mackinac of first consideration," Poel said of the report, which cited an ever-increasing shipping season and range of usefulness for the icebreaker.

The District's study examined 17 ports in Michigan and Wisconsin and immediately concluded that seven of the locations met a preliminary test of feasibility to moor the Big Mac. Ports that were disqualified included Sheboygan, Manitowoc, Green Bay, Menominee and Sturgeon Bay, Wis., and the Michigan ports of Presque Isle, Marquette, Ludington, Manistee, and Bay City.

The cities of Escanaba, Sault Ste. Marie, Traverse City, Muskegon, Grand Haven, Port Huron and Cheboygan passed the first test. Next, the Coast Guard set up requirements for acceptability for a moorage that included:

- A non-industrial dock or pier which has at least 21 feet of water alongside.
- Approaches that will allow the vessel to moor and unmoor without excessive difficulty or danger.
- A usable berthing space which will not endanger the ship due to other traffic in the vicinity or be substantially untenable due to weather or water action.

The Coast Guard rated the considered ports as follows in 1973:

- ESCANABA – Failed. Insufficient depth alongside the only acceptable dock.
- SAULT STE. MARIE – Failed. No acceptable dock and an undesirable 45-mile run through a channel to open Lake Huron.
- TRAVERSE CITY – Failed. No acceptable dock available.
- MUSKEGON – Failed. Available dock at Naval Reserve Center not acceptable due to heavy surge and wake from year-round car ferries that pass within 30 yards.
- PORT HURON – Failed. No acceptable mooring facilities available with the entire locale one of high traffic and high current.

- GRAND HAVEN – Retained. Moorings available at recently completed Corps of Engineers project with 21 feet alongside. The dock has an easy access approach from two directions and ample turning basin nearby in low traffic, low current, low surge area.
- CHEBOYGAN – Although this locale fails even the basic criteria for acceptability, it will be retained for comparative purposes since it is the current homeport of the Mackinaw.

It boiled down to Grand Haven and Cheboygan. Congressman Ruppe had the ship and Congressman Vander Jagt wanted it. The Coast Guard stated that "the only port meeting the preliminary tests of suitability, feasibility and acceptability is Grand Haven."

With battle lines drawn, the Grand Haven Tribune published a series of reports that compared the two communities, examining everything from location for convenience to storage space, major cities, shipping terminals and support for commissary supplies to weather studies, ice conditions, education and

The Mac creeps past Mission Point in the St. Mary's River.

morale considerations as well as parking for the ship. The newspaper's logic backed Vander Jagt's claims that if the ship could be located in Grand Haven it would be far more accessible to the many summer festivals it attends annually, though steaming time would be increased by about 16 hours for most ice-breaking operations.

"It has been suggested that moving the vessel southward would help in its civic responsibilities, meaning that it would be closer to the Cherry Festival, to the Blueberry Festival and closer to Fish Day," Ruppe told his fellow congressmen. "I suppose I am being insulting, but we are supposed to provide services for icebreaking and not services for the Blueberry Festival."

In the end, Ruppe convinced the House of Representatives that the Mackinaw's job was to break ice, and most of the hard water was nearer to Cheboygan, not Grand Haven. Vander Jagt lost on the roll call vote, 309 to 107, with the Michigan delegation dividing 12 to 6 against him.

The new moorings were funded and the ground-breaking was scheduled for Aug. 9, 1974. The event coincided with the second Mackinaw Alumni Reunion, held over a four-day weekend. The reunions began in 1969 and were held every five years through 2004 during the original Mackinaw's service life.

Edwin J. Roland, the icebreaker's first commanding officer, returned for the ceremony as a retired admiral. Admiral Roland participated along with Admiral O.W. Siler. Several of the Mackinaw's other former commanding officers were also able to attend.

At that first Mackinaw Alumni Reunion, held in 1969 to mark the silver anniversary of the ship, tragedy marred the celebration.

A group of 63 cadets were brought in from the Coast Guard Academy in New London, Conn., to train aboard the Mac that summer. The alumni group scheduled entertainment for the young men and set up a dance at East Elementary School to be held in conjunction with the alumni's activities including a softball game and a fireworks display. Young women from throughout the area were recruited to attend the dance with the cadets.

The day of the dance, Aug. 7, 1969, a group of cadets spontaneously went swimming in the Cheboygan River north of the moorings site after playing football in the warm sunshine. Cadet James B. Kehoe, a 19 year-old from Pittsburgh, Penn., drowned while swimming.

The dance was cancelled, the softball game

Adm. Edwin Roland attended the 30th Anniversary Alumni Reunion in 1974 and broke ground with Adm. O.W. Siler for the new Millard D. Olds memorial moorings. Added were a community building and upgraded storage, parking and recreation facilities.

moved to another field in town and the fireworks display moved from the site of the Coast Guard moorings.

When the community building was finished at the moorings site, the structure was dedicated to the memory of Cadet Kehoe. Family members of the young man came to Cheboygan to witness the placing of a plaque on the building, retained after it was remodeled in 2005.

The location of the ship proved to be the least of the Mackinaw's concerns in later years. Economic belt-tightening and outdated equipment on the ship caused a number of close calls in the 1980s and 1990s.

In 1982, congressional budget cuts and Coast Guard budget shortfalls began a series of decommissioning alarms that seemed to sound with increasing regularity. In January of that year the ship lay idle at Cheboygan while the Coast Guard began reviewing all operations to see where budget cuts could be made. In an effort to cut some $46 million in spending, the Coast Guard announced that it would be required to decommission ten cutters, close 15 search-and-rescue stations, and reduce operations at 16 other facilities. The Mackinaw was termed one of the "older and less efficient cutters" to be decommissioned. Meanwhile, the smaller "Bay Class" cutters worked overtime in a valiant effort to keep ships moving through the Straits of Mackinac, causing a public furor. When debate began about altering the status of the icebreaker, the first response came from U.S. Rep. Bob Davis, D-Mich., who just happened to be the ranking minority member on the House Committee on Merchant Marine and Fisheries.

Bob Davis

Davis knew that existing smaller icebreaking tugs couldn't do the Mackinaw's job, and that if the loss of the Mackinaw closed the lakes early or delayed the start-up of navigation there would be serious consequences for his state. The steel industry would have to either move Michigan and Minnesota iron ore by rail at greatly increased costs, or turn to overseas resources. The first option would reduce the domestic steel industry's ability to compete. The second alternative would be detrimental to the iron ranges of Michigan and Minnesota.

Michigan, home of more ports than all seven other Great Lakes states combined, stood to lose the most if the Mac was mothballed. Two bills were introduced before Congress to save the ship from retirement.

The Mackinaw embarks on another icebreaking mission in the 1970s. Note the bow scrapings from that year's icebreaking duties.

With Davis leading the fight in Washington, local community leaders went to bat for the ship in Cheboygan. City Manager William Chlopan formulated a plan to gather support for the icebreaker, and Cheboygan residents Larry and Ginny Hull were active leaders in a petition drive to save the ship.

"We distributed them to most of the businesses downtown and we got heaps and heaps of petitions back filled with signatures in just a few days' time," Ginny Hull told the Cheboygan Daily Tribune. Larry Hull served on the Mackinaw in the 1950s and 1960s and he and his wife remained active in Coast Guard Auxiliary activities for many, many years.

Modification of the cuts were finally authorized by Transportation Secretary Drew Lewis, who announced a "re-arrangement of priorities" and pledged that "better management and belt-tightening" had saved the Mighty Mac.

The pressure to save the ship worked, and the Mackinaw was saved. The ship's mission was redefined and the Mac was more or less "re-commissioned" with new sailing orders. Capt. Jim Honke took the cutter to the shipyard for a re-fit and returned with new equipment but a crew trimmed from 130 members to around 75. At least the ship was still based in Cheboygan.

However, the same rumblings were heard a few years later but summarily dismissed. In 1988, the topic arose again and seemed more serious than ever. The Coast Guard was facing a $105 million budget shortfall and planned serious cuts. The Mackinaw was on the chopping block again.

Cheboygan County Board of Commissioners Chairman Edmund "Pee Wee" Kwiatkowski again enlisted Davis' aid, only to learn that the budget cuts had been signed into law – the only question pertained to which vessels in the fleet would be decommissioned.

"We'll get out the petitions again," Chlopan told the Tribune, "and we'll do all we can to keep it here."

The Lake Carriers Association's powerful clout again reminded Congress and the Coast Guard that the Mackinaw was the only American icebreaker on the Great Lakes powerful enough to keep the navigation tracks open when a strong economy requires an early opening or an extension of the shipping season. The ship was also the only icebreaker capable of performing maximum ice management to prevent flooding in the St. Mary's River or the St. Clair River during high water levels.

In 1992, the howl about the Mackinaw's outdated systems, engines and difficulty obtaining replacement parts reached such a crescendo that the impetus towards decommissioning seemed based more on the Mac's obsolescence than it was about budgets, although once again money was the bottom line.

"The ship's broken, basically," the Mackinaw's Executive Officer, David Rokes, told WPBN-WTOM TV that year. "The battle is to finance fixing it. Someday it'll be more efficient to replace the Mackinaw. When that day comes, that will have to be some special ship."

Once again, through Davis' perseverance and the

The Mackinaw clears a track for the 1,000-footer Edgar B. Speer in the St. Mary's River.

support of the Cheboygan community, the ship got a reprieve. Major overhaul work was financed and completed and the cutter came out of the ordeal better than ever. Newer computer equipment replaced many obsolete systems involved in navigation, running the engines and electrical systems.

There was a feeling, though, that this might be the last "fix" for the Mac.

The Cheboygan community at last dedicated the facility that government tax dollars paid to build upon a site donated by a family who once operated a lumber business on the Cheboygan River. The dedication took place June 16, 2001.

Cmdr. Edward Sinclair presided over a ceremony that included a presentation of the genealogy of the Olds family. As of that date, there were 74 direct descendants of M. D. Olds and his wife, Leora Olds, and 64 of them were living at the time.

The Coast Guard Moorings, prior to the 2005 re-fit to accommodate two vessels, done by Ryba Marine of Cheboygan.

A 2005 dock extension brought the edge of the pier right out to the ship, eliminating an eight-foot gap that used to exist between the ship and the dock. Now the Mackinaw can rest in deep water right up against the pier.

The turning basin dock has also been remodeled with improved septic and electrical connections and dolphin piers to moor vessels of varying lengths. Landscaping completes a neat look at the site.

Millard D. Olds was born in Hartford, Mich., in 1860 and came to Cheboygan in 1892 to establish a staving mill. In 1904 he purchased the Clark and Nelson Saw Mill on the east side of the river. From there he built a railroad to serve the timber industry, active at the time, and from the neighboring forests moved the wood to the mills for manufacturing.

Olds was also involved in the lumber business in Oregon and in the sugar business in Paulding, Ohio.

It is on the site of the former saw mill that the present Coast Guard dock was established, improved, remodeled and expanded to what it is today – capable of supporting and serving many sizes and types of vessels and their shore-based needs, as well as buoy-tending operations and repairs.

The Mackinaw glides through Round Island Passage near the entrance to the Mackinac Island Harbor in 1971.

Red With Embarrassment?

A notable and controversial chapter of the Mackinaw's history took place when the ship received a new, red coat of paint in 1998. The original paint job when the ship was commissioned was a basic white, with the simple designation "W-83" on its sides appearing one year later in 1945.

The familiar and distinctive red slash or "racing stripe" did not appear on Coast Guard cutters, boats and aircraft until more recently in the Guard's history.

According to Coast Guard historian Florence Kern, the industrial design firm of Raymond Loewy/William Snaith, Inc. was hired to redesign the exterior and interior of the President John F. Kennedy's presidential plane in the early 1960s.

At the time, America's visual image had been neglected both inside as well as outside the U.S. Since image building played an important role in Kennedy's election, he approved their proposal for improving the world-wide visual identification of the U.S. Government. In 1964, the firm recommended that the Coast Guard adopt a symbol or mark that would be easily distinguished from other government agencies and easily applied to ships, boats, aircraft, stations, vehicles, signs and printed forms. Their design was a wide red bar to the right of a narrow blue bar, both slanted at 64 degrees. Centered on the red bar was a new emblem.

Studies were done with experimental markings for their impact on the public, as well as their long-run compatibility with the Coast Guard's mission and traditions. According to Kern, the reaction was overwhelmingly favorable. Three years later, on April 6, 1967, the now famous slash appeared throughout the Coast Guard.

In the final design, only the emblem changed. The traditional Coast Guard emblem was selected for centering on the red stripe over the new design.

The Mackinaw at Duluth, 2003.

In 1972, the U.S. Coast Guard cutter Glacier was painted red in order to improve visibility in Arctic regions. Commissioned May 27, 1955, Glacier was the free world's largest and most powerful icebreaker, capable of breaking ice up to 20 feet thick. Her Navy service extended to June 30, 1966, when she was transferred to the Coast Guard, under which she served until decommissioning in May, 1987.

Named for Glacier Bay off the Alaskan coast, the ship was painted red to contrast against the white ice fields it often worked. In the following year, 1973, all other icebreakers, except the Mackinaw, also were painted red.

The Glacier represented the "Glacier" class of icebreakers – a scaled-up version of the "Wind Class" – and had extended range, heavier ice-breaking capa-

The Mackinaw, painted red, approaches the dock in the Cheboygan River after an icebreaking mission in 2003.

bility and extended mission duration. The ship supported numerous polar scientific explorations, made several Antarctic landings and penetrations not previously accomplished, and performed a number of ship rescues.

In 1998, under the direction of Cmdr. K.R. Colwell, the Mackinaw also received a red coat of paint. Smaller cutters with icebreaking capability remained black with white and red trim. The move ended the distinction of the Mackinaw being the only icebreaker dressed in white.

It was indeed a shock to see the Mac for the first time in a totally different color. Like any ship undergoing such a change, it looked like a completely different vessel. This evoked plenty of passion among alumni of the Mackinaw, some who never saw the ship the same way again.

One of the Mac's former crewmembers, Jack Eckert, put many of his shipmates' feelings into perspective with a poem he published at "Jack's Joint," his popular column at the www.boatnerd.com Web site.

Eckert explained that the Mackinaw's mission was always within the Great Lakes, unable to go out and work in the salty seas. He felt that through all of the years from Buffalo to Duluth, Chicago to Detroit, Death's Door Passage to Whitefish Bay and on to Isle Royale the Mac was known as the famed "Great

Former Mackinaw officers at the 1974 Alumni Reunion included (l-r) Gordon Hall, John Bruce, Adm. Willard Smith, Larry Otto, Ben Chiswell, Frank Hilditch, Dwight Dexter and Clifford MacLean

White Mother" of the Great Lakes.

"Her beautiful white hull, in spite of her age, looked liked it was polished with auto wax," he wrote. "Today she is sporting the red colors of a harlot. What will she be called now, 'The Great Red Mother?' Hardly, that sounds almost Communist. Maybe 'Sweetwater Strumpet' would be more appropriate."

It is apparent that Eckert's feelings, like those of other old-timers, ran deep. Upon learning that this book would be published, Eckert sent the poem, asking that it be included in this chapter.

"Oh ye salt water scoundrels who perpetrated this abomination on to our lovely old lady – know this well – she needn't have a coat of red paint to cover up the rust of her long journeys and old age. An old song is recalled to mind – 'With Apologies.'

'For she's more to be pitied than censured
She's more to be helped than despised,
She's only a lassie who's ventured,
O'er the lakes' deep waters and pack ice.
Though by the dockside she waited,
She may yet mend her ways
That poor Great White Mother,
Who's seen better days."

Former Mackinaw captains at the 1984 Alumni Reunion included (l-r) Lawrence H. White, Capt. P.R. Taylor, Clifford R. MacLean, Gordon Hall, George D. Winstein and Adm. Martin Danielsen.

"It's quite a change to come here and see the ship painted red," said Renee Zimmer at the 2004 Alumni Reunion. "I know they said it's a safety thing, but the ship operated all those years safely while it was painted white, didn't it?"

Of course, those who joined the Coast Guard in more recent years gradually adjusted to the change in color and got used to it, especially if they had never seen the ship in white. Newcomers to the vessel only knew it as a red icebreaker, and every year that goes by produces more and more who served with red being the color of "their" ship. As the new Mackinaw ages, she will especially benefit from the effects of time since the newer ship was red from day one.

The last skipper of the original Mackinaw, Cmdr. Joe McGuiness, summarized his place in the ship's history by offering a message of safety and of moving on with progress. "I also like the idea of being visible out there in the white ice!"

A group of original Mackinaw Plank Owners at the 1984 Alumni Reunion posed with Capt. P.R. Taylor (2nd from right) including (from left) William Espe, Robert Marxen, George Jisa, Taylor and Norm Chastain.

Capt. Gordon Hall is recognized for his efforts in the Cheboygan community by Mayor Paul Lavigne at a 1980s alumni reunion.

Simon Antonie, an orginal plankowner of the Mackinaw, returned to Cheboygan for an alumni reunion in 1984.

Many alumni of the ship and Great Lakes boat watchers say they prefer to see the Mackinaw all dressed in white.

The Coast Guard ordered the Mackinaw painted red, like polar icebreakers, for consistency in purposes of visibility.

WAGB-83 – The Final Chapter

In the spring of 2004, the Cheboygan City Council sought citizen input to see if there were groups in the area who would be interested in taking over the U.S. Coast Guard cutter Mackinaw after the ship's decommissioning in 2006.

The Council approved a committee's recommendation that the city not be involved as an entity to take on the vessel's conversion to a museum and subsequent operation. If no one in Cheboygan could support the venture, then other communities would get their chance to bid on preserving the giant icebreaker.

"We're hoping that someone will see an opportunity here and help keep the Mackinaw in Cheboygan," Mayor James Muschell told the Cheboygan Daily Tribune. "The citizens deserve the chance to put together a plan to keep the ship here. Otherwise, we hope it will stay in the Straits Area."

Cheboygan Mayor James Muschell

At the center of the discussions were the questions of how many tourists would come to see the ship in Cheboygan if a suitable site could be arranged, and to determine if that number would be enough to pay the costly maintenance bills.

Examples of other Michigan maritime museums were cited, including the U.S. Coast Guard cutter Bramble in Port Huron, the museum ship Valley Camp in Sault Ste. Marie and the USS Silversides, a World War II-era submarine on display at Muskegon.

In August 2004, a meeting was held at the Cheboygan Opera House to determine a plan of action. The mayor named Jim Stevens, a retired entrepreneur with a business background as chairman of a blue-ribbon committee to keep the ship in Cheboygan.

Stevens was elected president of the volunteer task force, while former Mackinaw commanding officer Jim Honke became vice-president. Stephanie Jacobson and Roger Schwartz were voted as secretary and treasurer, respectively.

The committee assembled a business plan and established support from key civic and governmental groups as well as powerful political allies. The members explored various location options and visited other existing ship displays in Port Huron and in Kingston, Ont., Canada.

U.S. Rep. Bart Stupak

Bart Stupak, a Democratic U.S. congressman from Menominee, pledged support for the project as did U.S. Sen. Carl Levin, D-Mich., and U.S. Sen. Debbie Stabenow, D-Mich., after each received a joint resolution of support from Cheboygan County Commissioners and the Cheboygan City Council in October.

U.S. Sen. Carl Levin

According to Stevens, the organization hoped for a much easier task of converting the ship to its new use than some other preservation groups have had to endure.

U.S. Sen. Debbie Stabenow

"We will be way ahead of most museums because of the Mackinaw's size and the way it has been maintained," he continued. "The Mac will stay equipped as it is and be thoroughly maintained until the day it is turned over to us. The bridge, the engine room, all of the instruments and fixtures will stay onboard. The ship is even air-conditioned."

Stevens said that a dozen volunteers worked regularly on the project. Some wanted to re-paint the vessel white, to restore it to the look it had during the majority of its working life.

Meanwhile, the Mackinaw kept doing the job it had done so well for more than 60 years. The vessel broke ice, aided ships that were stuck in ice jams, and served as the hospitality queen of the Great Lakes as she'd always done.

"The ship runs better every day, it looks better every day and the crew gets more confident every day," Cmdr. Joe McGuiness said in July 2005. "It's like the old lady just isn't gonna give it up. You can throw the retirement party, but she doesn't want to quit."

McGuiness maintained the quality of work done by the Mackinaw through the last year the ship was in service, including one final season of icebreaking while the new Mackinaw trained "on the job." He spent a considerable amount of time working with his crew, getting them ready for their next assignment. Many among the crew knew the date they would cross the brow for the last time.

"There's no trepidation — we're all looking forward

Cmdr. Joseph C. McGuiness, the Mackinaw's last captain, breaking out a laker during 2004 icebreaking operations.

to the decommissioning," McGuiness said. "I'm trying to prepare them all for the next step in their careers."

After much work locally and in Washington, legislation sponsored by Stupak passed the U.S. House of Representatives to convey the ship to Cheboygan and to Cheboygan County. The Senate had to follow with its own version of the bill, championed by Levin and Stabenow.

"The conveyance of the cutter Mackinaw to Cheboygan is both a tribute to the ship that protected Michigan's waters and shores and cleared the ice paths for the nation's mariners," Stupak said. "This ship will now serve as an educational resource to help people better understand the history of the vessel, the Coast Guard, and the maritime history of the Great Lakes. In this role, it is imperative that Michigan keep this historic treasure."

Stupak said the Mackinaw had served the state of Michigan and the entire Great Lakes region for over 60 years, calling Cheboygan its only homeport.

"I see no better way to honor the life and name of the cutter than to retire it as a museum to its homeport in the Mackinac Straits area. It's a very historic resource — it breaks ice, it prevents flooding, it saves lives. This Coast Guard treasure will be a valuable cultural and educational benefit for generations to come."

The group planned to make the museum ship available for tours and educational ventures in an enclosed berth along the Cheboygan River.

The fantail deck became a giant picnic area during the 2004 Alumni Reunion Cruise.

The Mackinaw's crew on deck at quarters in 2005, while moored at the turning basin dock during repairs to the Cheboygan River moorings.

Cmdr. Joseph C. McGuiness conns the Mac on Lake Michigan during a fall work-up trip to get the crew ready for their last season of icebreaking in 2005. "This ship is a perpetual classroom," he said. "The training the crew gets here will be valauble to all of them later in their careers."

A volunteer blue-ribbon committee worked diligently in 2004-2005 to get the Mackinaw conveyed after decommissioning to the City of Cheboygan and Cheboygan County for use as a maritime museum.

This photo-illustration shows where the ship was planned for placement just north of the Cheboygan County Marina. The real Mackinaw is pictured at the dock, upper left.

The ship continues to display its stern homeport designation of Cheboygan, MI. The Icebreaker Mackinaw Maritime Museum group worked to keep the ship in Cheboygan, the vessel's only homeport.

A New Era, A New Mackinaw

Congressional committees spent at least 20 years analyzing how best to re-fit the giant icebreaker, how to extend its useful life and just what should go into a replacement vessel.

At last Congress approved the funding to build a new Great Lakes icebreaker, but not without a battle.

U.S. Rep. Bart Stupak, D-Mich., led a monumental campaign to get the new ship funded before the original version was decommissioned. He partnered with U.S. Rep. Dave Obey, D-Wis., and U.S. Rep Jim Oberstar, D-Minn., to create a blizzard of political pressure to get the job done.

"We did the heavy lifting," Stupak told the Great Lakes Seaway Review in 2005. "It would not have happened without the help of Obey and Oberstar."

Ironically, the congressmen's biggest ally may have been the weather. Particularly severe winters in 1992-93 and 1993-94 again proved the Mackinaw's worth to other voting politicians who began to understand the power of Great Lakes ice.

"All the lakes froze over that winter, even Superior," he said. "The Mackinaw is a legend on the Great Lakes and it's always been there when we need it. You can't replace a legend, but we're doing the next best thing."

Capt. Jonathan Nickerson was part of a design process to identify what was needed in a new vessel to keep the Great Lakes open during the long, frozen winters. The advent of new technologies all but guaranteed that any new version of the Mackinaw would be run with a far smaller crew than the numbers that sailed on the original vessel. Many systems were studied that had been successful in the polar regions as well as in Scandinavian ports where ice is thicker than in the Great Lakes.

Another consideration sure to be incorporated into the new icebreaker's design would be the ability for multiple-mission capability. Coast Guard budget cuts had removed some smaller ice-breaking tugs and buoy tenders from service, and the new ship could be designed to handle some of their duties if properly equipped.

In tough economic times, the importance of building a ship that could fulfill all roles was paramount to replacing the aging Mackinaw that had done its job so well for so long. The original version worked, but modern technology would have to succeed in getting the same job done through its own means, with fewer crewmen and handle other duties besides.

The new ship, also named Mackinaw, was

planned as a one-of-a-kind icebreaker in the Coast Guard's fleet – 240 feet long, with a beam of 58 feet, and a 16-foot draft.

Launched April 2, 2005 at Marinette, Wis., into the Menominee River, the ship was christened by Jean Hastert, wife of Speaker of the House Dennis Hastert. Many Cheboygan citizens drove to see the launch at Marinette Marine Corporation and others chartered a bus to witness the event.

The ship's design uses two fully azimuthing podded propulsors, meaning its "podded" — or protected — propellers can rotate 360 degrees for greater maneuverability. It is the first such vessel in the Coast Guard fleet and for that matter, the first for the U.S. government. The azipods deliver a combined 9,200 horsepower. Mackinaw also has a 500 horsepower tunnel thruster forward.

The new ship is able to break 32 inches of level ice at three knots ahead or two knots astern and eight to 12 feet of brash ice — chunks of ice that have refrozen — at three knots ahead or two knots astern.

Topside, Mackinaw's crew enjoys the most mod-

The new Mackinaw WLBB-30 under construction in 2004 at Marinette Marine Corporation's huge indoor shipbuilding facility.

The new Mackinaw WLBB-30 was launched April 2, 2005 at Marinette Marine Corp. into the Menominee River.

ern bridge equipment in the fleet. The new cutter has a fully integrated bridge system combining the control-and-navigation system, voyage-planning system, dynamic-positioning system, an electronic chart- display-and-information system and radar. The vessel is equipped with a state-of-the-art communications suite. The bridge is designed for operation by one person in unrestricted waters, although plans are to operate with two or more.

The new Mackinaw, commissioned as WLBB-30, also has an astern conning station located aft on the deckhouse to better monitor operations astern. The 50-person crew — 25 fewer than assigned to the current icebreaker — is literally able to maintain the new ship's position within a 10-meter circle in up to 30-knot winds and up to 8-foot waves, with a margin of error near zero.

The bridge is configured with an area on the port side aft to accommodate mission command-and-control activities for such things as search and rescue and homeland security, while allowing for the safe navigation and ship-control functions forward. The new Mackinaw also has port and starboard bridge control consoles.

One of the ship's missions will be buoy tending.

In addition to icebreaking, one of WLBB-30's primary missions is servicing aids to navigation. The vessel has 3,200 square feet of buoy deck space, configured exactly the same and with the same equipment as the large Coast Guard buoy tenders. Mackinaw is able to service buoys in water as shallow as 18 feet and its crane is able to recover buoys, chain, and sinkers weighing up to 20 tons.

Capt. Don Triner was named as the vessel's first commanding officer and oversaw pre-commissioning details including the building of the ship, the sea trials and installation of the first crew as well as its delivery to Cheboygan. An enthusiastic and personable man, Triner made sure that the original

Capt. Donald Triner, first commanding officer of the new U.S. Coast Guard Cutter Mackinaw, WLBB-30.

Mackinaw and its crew received all the respect and honor that was rightly due during the winter of 2005-2006 when the two vessels broke ice together. The old Mac was planned for decommissioning and the new ship to be commissioned June 9, 2006 in a ceremony of farewell and welcome with duties transferred.

Although many veterans of the original Mackinaw doubted the ability of the new cutter to perform icebreaking operations with the same results, the new Mackinaw's design encompassed a new way of doing the job and the capability of doing other jobs too.

A new queen may have been crowned in the Great Lakes, but the royalty, history and stories of the original ship will live on in the hearts of those who fondly remember the original Big Mac.

The bridge on the new Mackinaw utilizes computer-driven systems for steering, propulsion, navigation and other shipboard functions.

It appears that the ship can practically reverse course into its own wake at 15 knots, thanks to its azimuthing propulsion pods that can rotate 360 degrees.

New Mackinaw underway during sea trials in September, 2005.

IN MEMORIAM

Jack Eckert

1939-2005

This book could not have been written in the same way that it was without the encouragement and help of Jack Eckert. A former Mackinaw crewmember, Jack was a major motivator for the publication of this book through his fact-finding, persistence for detail and his humor. He also had a heart of gold when it came to anything that would tell the story of a ship he truly loved, the U.S. Coast Guard cutter Mackinaw.

Sadly, he did not live to see the fruition of his helpful ways. Jack died in 2005. May he sail smooth seas under sunny skies, along with his fellow shipmates who have also crossed the bar.

ACKNOWLEDGEMENTS

Like the efficient operation of a ship, it takes a good team to put together a book like this one about the U.S. Coast Guard cutter Mackinaw.

Besides the tremendous inspiration of the late Gordon Turner, I owe a huge debt of thanks to my employer - the Cheboygan Daily Tribune. Despite covering the Mackinaw for television and radio for many years, there was no better resource possible than the Tribune's archives. Whenever there was a question about a date, a person or an occurrence, a trip to the newspaper's "morgue" usually brought about an answer to the question and provoked another topic for consideration in this book. There was no question from the very beginning that I had the backing of Publisher Valerie Rose and Editor Rich Adams, who each provided encouragement and advice as did fellow Tribune staffers Shawna Jankoviak and Erica Kolaski.

The patience of composition editor Dan Pavwoski and the creative brilliance of graphic design artist Charles Borowicz allowed my concept of what this book could be to become reality. My co-workers, without fail, lent expertise, grammar and common sense when it was needed.

The wonderful photographs of so many boat-watchers who also love the Mackinaw were key components of placing the Mac at festivals, in various Great Lakes ports and on the job at sea. The cooperation of newspapers like the Grand Haven Tribune, the Holland Sentinel, The Toledo Blade and the Alpena News also lent perspective, history and detail to this project.

But the story of the Mackinaw is as much about the people who captained her, sailed her, funded her and fought to keep her in Cheboygan as it is about an extra-thick hull or a 12-foot diameter bow propeller.

Time and time again, the people associated with this great ship answered the call when I came looking for stories, verification or names from more than 60 years of service. How else do you identify a piece of equipment in the No. 2 engine space than from someone who served on the ship?

I could not have predicted the number of times I had to ask Ed Pyrzynski to look at one more batch of photos, but he greeted each inquiry with a smile. What a treat it was to meet so many people in my own community - and new friends from across the country - that had served about the "Great White Mother." Just when I thought I had reached a dead end looking for a picture or a name relating to a particular aspect of the Mackinaw, I would get a package of photos from Jack LaLonde or Bill Lorenz, an e-mail from Gary Williams or a visit from Don Wright, Bill Tomak or Jim Honke.

They gave me a sense of direction and realism that helped me understand the lives of those associated with this giant icebreaker. When Leo Cocciarelli said he had a photo of a freighter's bow stuck in the stern of the Mackinaw, you just knew that must have been quite a day onboard the ship. Those are the stories I wanted to bring to life.

The access and answers I received from public affairs officers James Connors and Elizabeth Newton, and from the Mackinaw's last crew were invaluable.

I am also appreciative of the encouragement I received from Cmdr. Joe McGuiness, the Mac's last skipper, and Jim Stevens, who led a courageous campaign to keep the Mackinaw in Cheboygan as a museum ship after decommissioning.

Above all, my thanks and love to my wife, Karen, for her grammar correction, encouragement and advice on a project that seemed to only get bigger and bigger with time.

Mike Fornes - 2005

PHOTO CREDITS

The author gratefully acknowledges the following photo credits and is appreciative of their generous cooperation in contributing to this book:

PHOTO CREDIT	PAGE NUMBER(S)
Albion College Archives	2
Rogers City Historical Museum	2
Cheboygan County Museum/ Cheboygan Daily Tribune	30, 31, 32, 33, 38, 39, 40, 44, 54
Cheboygan Daily Tribune	I, IV, V, 10, 12, 13, 33, 83, 129
Ellis Olson – 3, 5, 13, 33, 135.	
Jack Eckert - www.jacksjoint.com	46, 66, In Memoriam
Ken Rocheleau	17, 24, 26, 139
Joe Etienne	25
Jerry Pond	84
U.S. Coast Guard	24, 29, 47, 63, 68, 85, 114, 119, 128, 158
Gary Williams	18, 19, 42, 70, 72, 73, 96, 99
Michigan Shipwreck Research Associates	79
Robert Packo	125
James Hebert	76, 77, 96, 102, 110
Neil Schultheiss www.boatnerd.com	Front/back covers, 74, 75, 76, 77, 78, 96, 100, 108, 109, 120, 126, 128.
Ron Prater	128
Shirley Ringo	65
Todd Davidson	135
Andy Loree	118
Jim Hoffman	47, 146
Barbara Algenstedt	61

PHOTO CREDIT	PAGE NUMBER(S)
Peter Wyatt	113, 123, 124, 133, 157
Jack LaLonde	15, 16, 17, 20, 22, 32, 37, 39, 41, 45, 59, 64, 71, 81, 86, 92, 93, 94
Bill Lorenz	24, 27, 39, 45, 51, 62, 67, 69, 82, 90, 91, 97, 122
Ed Pyrzynski	27, 31, 32, 34, 36, 45, 50, 54, 142, 143, 144, 145
Don Wright	33, 49, 56, 69
Richard Hess	58
Leo Cocciarelli	21
Michigan Congressional Library	131, 132, 149
Sault Evening News	1, 126, 127
Duluth Shipping News - Ken Newhams	130, 140
The Associated Press	23, 136
USCGC Mackinaw Archives	4, 5, 6, 7, 9, 29, 31, 34, 35, 43, 86, 89, 98, 102, 104, 105, 106, 107, 112, 121, 134, 149
New USCGC Mackinaw Photographers: EMC Herbert Boggs, IT1 Stephen Chipman, SA Sara Miller	156, 158, 159, 160
Photo Illustrations by Charles Borowicz	Front/Back Covers, II, 153
Author's photographs	3, 52, 53, 55, 71, 88, 95, 100, 101, 103, 111, 115, 116, 137, 138, 141, 147, 148, 150, 151, 152, 153, 154

ABOUT THE AUTHOR

Mike Fornes

Mike Fornes has covered the U.S. Coast Guard cutter Mackinaw for several media outlets in Northern Michigan, including radio and television stations and the Cheboygan Daily Tribune.

A resident of Mackinaw City, Mich., he enjoys scuba diving and sailing among his interests in the Straits of Mackinac area.

He is frequently in demand as a guest speaker and presenter to tour groups, cruise ship organizations and historical societies.

He is also the author of "Mackinac Bridge – A 50-Year Chronicle 1957-2007" to be published in 2007.